Cultural Sociology

Cultural sociology is widely acknowledged as one of the most vibrant areas of inquiry in the social sciences across the world today. The Palgrave Macmillan Series in Cultural Sociology is dedicated to the proposition that deep meanings make a profound difference in social life. Culture is not simply the glue that holds society together, a crutch for the weak, or a mystifying ideology that conceals power. Nor is it just practical knowledge, dry schemas, or know how. The series demonstrates how shared and circulating patterns of meaning actively and inescapably penetrate the social. Through codes and myths, narratives and icons, rituals and representations, these culture structures drive human action, inspire social movements, direct and build institutions, and so come to shape history. The series takes its lead from the cultural turn in the humanities, but insists on rigorous social science methods and aims at empirical explanations. Contributions engage in thick interpretations but also account for behavioral outcomes. They develop cultural theory but also deploy middle-range tools to challenge reductionist understandings of how the world actually works. In so doing, the books in this series embody the spirit of cultural sociology as an intellectual enterprise.

Bernhard Giesen • Werner Binder
Kim-Claude Meyer • Francis Le Maitre

Human Societies

An Evolutionary Introduction to Cultural Sociology

palgrave
macmillan

Bernhard Giesen
University of Konstanz
Konstanz, Baden-Württemberg
Germany

Werner Binder
Dept of Sociology
Masaryk University
Brno, Czech Republic

Kim-Claude Meyer
Ministry of Education,
Children, Youth
Luxembourg City, Luxembourg

Francis Le Maitre
DHBW
Villingen-Schwenningen
Germany

ISSN 2946-3572 ISSN 2946-3580 (electronic)
Cultural Sociology
ISBN 978-3-032-07696-0 ISBN 978-3-032-07697-7 (eBook)
https://doi.org/10.1007/978-3-032-07697-7

Cover Credit: Orbon Alija/Getty Images

This Palgrave Macmillan imprint is published by the registered company Springer Nature Switzerland AG.
The registered company address is: Gewerbestrasse 11, 6330 Cham, Switzerland

No one simply disappears
In memoriam
Bernhard Giesen
1948–2020

Series Editors' Preface

Civilizational analysis covers large sweeps of human history. It surveys vast continents and myriad cultures, looking from a distance and attempting to make sense of it all. It is a genre that presents a challenge to even the foremost intellectuals. We think of giants such as Herodotus, Pliny, Montesquieu, or more recently Halford Mackinder and Arnold J. Toynbee. Max Weber and Karl Marx come to mind for the sociological audience. For the humanist there might be Jacob Bronowski, whose remarkable television series The Ascent of Man moved effortlessly between the cave paintings of Lascaux and the game theoretical mathematics of John von Neumann. Yet today, such activity has become decidedly unfashionable. Michael Mann and Robert Bellah have attempted to chart the sweeping path from band societies via Antiquity to the foundations of Axial religion and the modern state. But they are marked exceptions. We live in a world where academic caution, methodological prowess, and niche specialism are rewarded. Progress in ideas, to paraphrase Prufrock, is now measured out in coffee spoons.

The task is doubly burdened. For aside from being out of step with the times, it requires a very rare skill set. In a strange way this is an unprofessional one. The composer Richard Wagner was described by Thomas Mann as a dilettante. He could turn his hand to anything—beyond composing music there was stage craft, architecture, stage design, dramaturgy, conducting, writing, and philosophizing. Mann was an admirer and intended his characterization as a compliment. Wagner's creative genius came from an inquisitive, enthusiastic dabbling in many things. Boundaries could be crossed, boundaries that locked others down into the treadmill patterns of thought of professional expectation.

Likewise, the civilizational scholar must have a remarkable non-specialist knowledge of technology, geography and climate, social organization, culture, and economic life. But this surfeit must be channeled through an analytic mind that abstracts common patterns rather than listing curiosities and exceptions. Outcomes can no longer be reduced to the "local" or the "contingent"; the civilizational analyst sees deep resemblances in process and form that have repeated time and time again over history. For example, the horse had similar impacts on warfare and empire in the plains of Mongolia as it did with the tribes of the American West after it arrived with the conquistadores, and then escaped. Bureaucracy in Ancient China and the British Civil Service measured talent with exams and helped build powerful, high trust regimes. The stirrup and the plough engendered phase shifts much like the steam engine and the computer.

Whereas specialists on Ancient Rome or the Aztecs will protest that the object of their particular study is unique and special, or that we need to master the language to truly understand them "from within," the civilizational analyst (dilettante if you wish) will only see deep continuities in military organization and war, slavery, the impacts of writing on power, or the forms of pantheistic religion. At the end of the day, Quetzalcoatl is just another Zeus, Ra, Jupiter, or Odin. To be able to see these kinds of similarities can do violence to particularity, but it can also generate insights of the first order that push us toward a cumulative, empirically grounded scientific program of social explanation on the grandest of scales.

But even as it cumulates "scientifically," this mode of inquiry requires a kind of double consciousness. There has to be an anthropological detachment from judgment for civilizational work to be done properly. In a sense, we must marvel or at least respect the accomplishment of social order within each civilization, even if we disagree with what we see. The details of the history of that social order are often unpleasant—torture and slavery, serfdom, colonialism, destinies ascribed by the vagaries of birth, and custom. But we need to study how it all works, not turn away. Findings that challenge our contemporary common sense about what people should be or how they should live must at the end of the day be respected as social facts to be studied. We cannot just pretend people were "wrong," or that deep down they were just like us in their dreams and values.

And so it is that with the lectures that make up this book. Bernhard Giesen found a way to master a vaunted but difficult terrain. The editors' introduction maps out just how Giesen does it. Here we indicate two themes that seem to be of particular interest to ourselves.

One is the way that Giesen moves beyond Mary Douglas, perhaps the figure who has done the most in contemporary anthropology to engage in thinking civilizationally and relativistically on matters relating to society and culture. The great anthropologist famously insisted that forms of social organization (e.g., solidary groups, hierarchies, boundaries) were universal building blocks and that these generated predictable continuities in thought styles, problem-solving, and justification. The result could be breathtaking insights through unexpected comparison—a market trader in an African village and a Wall Street banker might have similar views on how to manage a dispute or best distribute scarce resources. This is anthropological relativism at its best. But Douglas did not have a big picture theory of modernity, of societal evolution, or of how scale, technology, communication at a distance, moral evolution, and societal capacity might have made such comparisons about "groups" problematic. With his Weberian instincts, Giesen saw the need for a more complete answer to the question "How did we get here?" even as Douglas was still asking, "Did we really get anywhere?"

The other advance Giesen made was in drawing on symbolic anthropology—from Durkheim and Levi-Strauss to Victor Turner and Clifford Geertz—to demonstrate the role of arbitrary cultural codes and grammars in shaping life. If a pollution belief was universal, what it latched onto in the "here and then" was contingent. What was sacred and profane, valued and cursed was to some extent not that predictable at all. Perhaps, Giesen recognized that there was a place for the "local" in civilizational analysis after all. Making the cultural turn, he had become a pioneer in the new field of cultural sociology. Not fully trusting herself to "lean in," Douglas remained a cultural anthropologist.

Bernhard Giesen was one of the most brilliant sociological theorists of our day. It is an extraordinary achievement that his former students, having discovered these unpublished lectures, have engaged in the scholarly effort to edit and introduce them. It is an honor and a privilege for the Palgrave Series in Cultural Sociology to bring them to the light of day.

New Haven, CT, USA

Philip Smith
Jeffrey C. Alexander

Competing Interests The author has no competing interests to declare that are relevant to the content of this manuscript.

In Gratitude

It was during the summer semester of 2004 that we first encountered Bernd Giesen at the University of Konstanz. His introductory lecture "Kultursoziologie I" (Cultural Sociology I) became the intellectual seedbed for what would later become the *Human Societies* manuscript. Written in 2006, the manuscript exhibited a remarkable level of conceptual maturity, arguably too mature to remain unpublished for so long. It lingered in the background, awaiting the moment in which its symbolic potential might be fully realized.

And yet, no manuscript, however refined, and no introduction, however precise, can truly recreate what was most distinctive about that course: the performative presence of Bernd Giesen in front of his students. His lectures were not mere deliveries of knowledge; they were acts of cultural communication—symbolic performances that embodied charisma and intellectual play. For Giesen, a good lecture was not a monologue but a dynamic exchange. He explained and provoked, raised questions, and distributed symbolic rewards, "bonus points," for moments of critical insight. What emerged in these encounters was that rare sociological energy which C. Wright Mills once called the "sociological imagination": the capacity to perceive biography and history, structure and agency, within a single interpretive frame.

We hope that this manuscript offers at least a distant echo of that pedagogical moment, that it conveys, however partially, the sense in which culture appeared not merely as a theory, but as an unfolding experience of meaning.

We are especially grateful to Christiane Giesen, who agreed without hesitation and with great enthusiasm to the publication of this manuscript. Her consent was more than a gesture; it was an act of cultural transmission, ensuring that a symbolic world, once enacted in a specific time and place, could reach beyond its original context.

Francis Le Maitre and Kim-Claude Meyer
In the summer of 2025

Contents

EDITORS' INTRODUCTION BY WERNER BINDER

Bernhard Giesen's (r)Evolutionary Introduction to Cultural Sociology: An Extraordinary Perspective on Human Societies

Werner Binder

Kein Wesen kann zu nichts zerfallen
(No being can to nothing fall)[1]

"Nichts verschwindet einfach." *Nothing simply disappears.* This has been the unofficial motto of Bernhard Giesen's introductory lecture "Kultursoziologie I" (Cultural Sociology I), which he taught every year at the University of Konstanz. Officially, Giesen held the chair for macrosociology, like he did previously at the University of Giessen (yes, professor Giesen from Giessen), but by the time he came to Konstanz in 1999, he had already become a full-fledged cultural sociologist. It is this introduction to cultural sociology, refined over years of teaching, that forms the basis of the English manuscript *Human Societies.* Written down in 2006, it served as a script for a series of lectures he gave in the Spring semester 2008 in New Haven at Yale University. Giesen's lectures left their imprint

[1] Opening line of Johann Wolfgang von Goethe poem "Das Vermächtnis" (translation by Mildred Harnack, an American who was executed as a member of the German resistance by the Nazis).

B. Giesen et al., *Human Societies*, Cultural Sociology,
https://doi.org/10.1007/978-3-032-07697-7_1

on students on both sides of the Atlantic, but his deteriorating health condition and untimely death in 2020 upended any plans he might have had for a revision and publication of the manuscript.

Giesen was an exceptionally gifted performer. Far from being dry and orderly, he did not conform to the stereotype of a German professor—at least not in every respect. He talked freely and passionately, not shying away from controversial statements. For Giesen, being a cultural sociologist not only meant putting meaning first but also staying clear of any form of moralism or political correctness. In his view, the practice of slavery in ancient Greece (p. 103) was no less "natural" than the idea of human rights that emerged with the modern nation state (pp. 52f., 180). As a modern citizen, Giesen condemned slavery; as a cultural sociologist, he understood the practice of slavery and his own condemnation of it as a consequence of different classification systems. At times, his lectures may have provoked the moral sensibilities of his audience, but they were always thought-provoking. For his Yale lectures, he not only hand-picked photographs and showed them on an old-fashioned projector, but also borrowed artifacts from the university repositories as props for his performance. At home and abroad, he offered cultural sociology the stage it needed to inspire students and young academics. Giesen was a natural performer in the lecture hall but we can find similar qualities in his writings too. Or in the words of his longtime friend and collaborator, Jeffrey C. Alexander:

> His writing was almost freakishly creative, leaping with daring confidence toward the unknown; yet, it was also tightly argued. If his theorizing was unapologetically abstract, his empirical discussions were thick with the texture and stuff of social life, the illustrative, the everyday, the empirically concrete. (Alexander 2021: 238)

The script for *Human Societies* reflects Giesen's virtues and idiosyncrasies as an original thinker. It offers a sweeping view over the sociocultural evolution of the human species, which is nevertheless rooted in Giesen's profound knowledge and understanding of history. Arguably, there are only few scholars, dead or alive, with the chutzpa and the skill set necessary for such a grand endeavor. Contemporary sociologists have largely retreated from the ambition to paint a "big picture" of human history. Outside the discipline, meanwhile, books like *Sapiens* by the historian Yuval Noah Harari (2014) or *The Dawn of Everything* by the anthropologist David Graeber and the archeologist David Wengrow (2021) have been wildly successful. Arguably, *Human Societie*s is unlikely to become the next blockbuster as it lacks a simplistic narrative à la Harari nor does it promote a popular political agenda like Graeber and Wengrow. Instead, Giesen sketches a complex and multilayered panorama of humanity from the perspective of a curious yet detached observer. In his reconstruction of human history from a cultural sociological perspective, Giesen was indeed "hunting a big animal"—to use a phrase he himself employed frequently to compliment the bold projects of others. Yet, *Human Societies* is not only Giesen's most ambitious, but also most accessible work. He decided to forgo the lengthy theoretical exposition which is often found in his other works. Still, the entire manuscript is animated by an intellectual spirit clearly distilled from the fruits of decades of theoretical reflection. Make no mistake: *Human Societies* may be written lightly, but it should be taken seriously.

In the context of contemporary scholarship, *Human Societies* remains a peculiar work, with only few references to other works—and mostly from the anthropological and sociological canon. While Giesen was incredibly well-read, he was not exactly known for citing all his sources and influences. Well-aware of the fact that he stood on the shoulders of giants, his foremost concern was the argument he presented, which had to stand on its own feet. Compared to his other books, it is rather remarkable that he chose to quote other authors verbatim in *Human Societies.* This has partially to do with its pedagogical character as a lecture script, for example, citing historical examples to add empirical texture or introducing sociological key concepts such as Weber's charismatic (p. 54) and traditional (p. 114f.) types of authority. More important, however, are the quotes that directly contribute to the unique character of Giesen's argument.

Arguably, the most fascinating use case for quotes in *Human Societies* are what can be called "historical flashforwards"—to use a cinematic term for "flashbacks" from future events. Examples are an excerpt from a text on the behavior of passengers on a train, which Giesen uses to illustrate classification in simple societies (p. 48), or a quote from Goffman that showcases the social consequences of the rise of urbanity in conquest societies (p. 106f.). In addition, Giesen offers his own examples as "flashforwards," for example, when he illustrates the logic of gift-giving with Christmas gifts (pp. 49, 65) or compares trophy-taking in early societies to sports trophies and touristic souvenirs (p. 96). In doing so, Giesen engages in a double movement, which complements the classical Durkheimian strategy of using anthropological and historical sources to understand contemporary society by a reverse approach in which historically distant phenomena are viewed through a contemporary lens. Thus, *Human Societies* advocates a cultural-sociological understanding of human history and contemporary society, rendering the former familiar and the latter strange. As peculiar as these "flashforwards" might at first seem, they are intimately connected to Giesen's unique understanding of sociocultural evolution.

Nothing simply disappears. This dictum captures Giesen's unique evolutionary approach to human societies. For him, there was no sharp distinction between simple and modern societies, between the hunting of big animals and the chase after big ideas. Unlike in biological evolution, nothing is really lost in sociocultural evolution, at least when it comes to social and cultural forms such as ritual or gift-exchange. In terms of intellectual style, Giesen was more Durkheimian than Weberian, but Simmel was also a major influence: In his last book, *Zwischenlagen* (2010), he made extensive use of the form of the essay, whereas in *Human Societies*, he offers a variation on Simmel's formal sociology, positing the existence of relatively stable social and cultural forms, which are independent of their specific historical content. For Giesen, the general form of the ritual remains the same throughout its specific historical instantiations and variations. In terms of formal characteristics, rituals today are not that different from the rituals of hunter-gatherers. While Giesen's thinking was profoundly historical, he was not afraid to generalize. At the same time, he rejected overly simplistic and linear narratives of human progress, including the conception of evolutionary stages:

> Thus, the evolutionary perspective of this book should NOT be mistaken as a neat historical sequence of evolutionary stages that succeed one another and replace the preceding patterns entirely. There is no neat separation between the so called primitive culture and the modern mind—as many early anthropologists assumed. (p. 42)

Nowadays, the rejection of evolutionary stages seems hardly revolutionary (cf. Graeber and Wengrow 2021). What sets Giesen apart from contemporary critics is his continued defense of evolutionary thinking, in line with the work of his mentor Shmuel Eisenstadt (1982) and of the late Robert N. Bellah (2011). Refusing to throw out the baby with the bathwater, Giesen conceptualizes sociocultural evolution as a process in which new classification systems, new forms of social organization, and new cultural practices appear without the old forms simply vanishing. Classification systems centered on the distinction between the sacred and the profane, forms of domination like charismatic authority, and cultural practices such as gift-exchange, which according to Giesen all originated in the earliest human societies, continue to live on and shape contemporary society. When new social and cultural forms appear, they tend to take the center stage of society displacing older forms in the process. Yet, these older forms do not simply vanish but have a continued existence at the periphery of society: "All of them remain with us" (p. 42).[2]

This unique approach to human societies can be captured by Ernst Bloch's formula "Ungleichzeitigkeit des Gleichzeitigen," elsewhere translated by Giesen as "noncontemporaneity" and defined as the "coexistence of phenomena that we relate to different historical periods or attribute to different evolutionary stages" (2004: 28). Critical of Bloch, a modernist progressivist for whom noncontemporaneity was pathological and potentially dangerous, Giesen normalized the concept and gave it a postmodern bend:

[2] Arguably, the insight that "nothing is ever lost" can also be found in Robert N. Bellah's scholarly treatise *Religion in Human Evolution* (2011: 13), where it is called "one of the central principles of this inquiry" (p. 267) and even "a premise of this book" (p. 489):

> Nothing is ever lost. Just as the face-to-face rituals of tribal society continue in disguised form among us, so the unity of political and religious power, the archaic "mortgage," as Voegelin called it, reappears continually in societies that have experienced the axial "breakthrough." Kings who ruled "by divine right," are obvious examples, but so are presidents who claim to act in accordance with a "higher power." (Bellah 2011: 267)

Given the fact that Giesen's unpublished manuscript was completed before Bellah's book, it is best to assume that both authors arrived at a similar conclusion through a process of convergent intellectual evolution. Yet, both seem to have taken very different paths: Bellah (2011: 13) introduces his principle through an analogy with the developmental psychology of Piaget and others; for Giesen, the conclusion that "nothing simply disappears" seems to be premised on his peculiar understanding of sociocultural evolution (1991a, 1991b). Arguably, despite being proclaimed a central principle, Bellah doesn't make much use of this insight throughout the book. It seems more of an afterthought, which—perhaps—came too late to be incorporated in a more comprehensive and effective manner. Also, as a premise it is largely left unaccounted for. Bellah does not tell us why "nothing is ever lost"—neither in his *Religion in Human Evolution* nor in one of his other works. In comparison, Giesen's "nothing simply disappears" offers several advantages: It is not confined to "religious history" (Bellah 2011:13), but applies to sociocultural evolution in general; it is implemented more effectively, largely due to Giesen's innovative use of 'flashforwards'; last but not least, while lacking an explicit justification in *Human Societies*, it follows from Giesen's conception of sociocultural evolution developed in other works (1991a, 1991b).

> If we reverse the perspective and consider the coexistence of historically heterogeneous phenomena to be the normal situation we can account for translations, bricolage, syncretism, but we lose our phenomenon—Bloch's *Ungleichzeitigkeit des Gleichzeitigen* as a special phenomenon becomes invisible. (Giesen 2004: 29)

For Giesen, contemporary society is characterized by noncontemporaneity on a fundamental level, being not only a bricolage of distinct cultures but also a blend of different historical epochs. Yet, he never fully embraced postmodern thinking but remained committed to the quest of discerning order in history. Following Weber, Giesen thought of such an order as ideal-typical, a theoretical abstraction from an otherwise messy historical reality:

> The evolutionary perspective used here is therefore more a presentation of ideal types than an accurate description of historical processes. Throughout the course of history, real societies have frequently blended these ideal types. For example, a society may have been in most respects a conquest society while also demonstrating some features typical of empires. (p. 41f.)

In *Human Societies*, Giesen distinguishes between six ideal types of societies, from the hunter-gatherer band to the modern nation state, each type being characterized in terms of four distinct types of modes, namely production, classification, circulation, and representation. Interestingly, Giesen himself blends these ideal types to make sense of broader social formations, which do not sit comfortably within the general scheme. The best example is probably Giesen's characterization of European colonial empires (p. 146), where the imperial distinction between center and periphery is coupled with the strict boundaries of a territorial or nation state. We will return to the different types and modes that structure *Human Societies* in detail after a brief reconstruction of the evolution of Giesen's thinking leading up to it.

From Macro Sociology to Cultural Sociology: The Backstory of *Human Societies*

If nothing simply disappears, maybe also the converse is true: nothing simply appears out of nowhere. "Every theory has its backstory," Giesen (1991b: 12, our translation) states in his *Entdinglichung des Sozialen*, a

German monograph that not only offers an evolutionary perspective on postmodernity but also marks a turning point in his own thinking, highlighting for the first time "culture" as a categorical foundation of social action and human society. While highly original, Giesen's understanding of culture and history in *Human Societies* didn't come from nowhere. It is as much a product of his intellectual biography and influences as it is his creative work as an original thinker. As already mentioned: Giesen was exceptionally well-read, yet often reluctant to share his sources and influences. In the following, we would like to reconstruct parts of the backstory of *Human Societies*, not to relativize the novelty of his thinking, but to allow for a historical contextualization and deeper appreciation of his work.

The story starts in Heidelberg, where Giesen was studying history and sociology (receiving his M.A. in 1972) during the tenure of the famous German historian Reinhart Koselleck (who left in 1973 to Bielefeld). Koselleck had a tremendous influence on Giesen (as he readily admitted), not only on his performative style but also his later research interests. The discussion of the rise of the bourgeoisie, the role of intellectuals and the birth of the modern nation state in *Human Societies* are deeply informed by Koselleck, despite the fact that he is never explicitly mentioned or referenced. Yet, even when Giesen was under the influence of others, he always gave their concepts and theories his own twist, using them to satisfy his own sociological curiosity and for his own intellectual ends. Koselleck, for example, coined the term "saddle time" (*Sattelzeit*) for the period between 1750 and 1850, roughly the transition between the territorial states of early modernity and the modern nation state. While Koselleck later disavowed the term, Giesen (1998/1993) re-interpreted it as a modern "axial age," in which the nation state was forged by peripheral intellectuals. While Koselleck left his intellectual imprint on Giesen, the fresh graduate left Heidelberg for Augsburg and then Münster to pursue entirely different research interests.

Giesen's Early Works: A Macro-Sociological Perspective on **Human Societies**

In the 1970s, Giesen primarily worked on the methodology and theory of social sciences regarding himself a Popperian. In 1974, he received a PhD from the University of Augsburg with a thesis on theories of structural inconsistence, after which he moved on to Münster, where he received his postdoctoral degree in 1980 with a work on societal identity and evolution. The same year, the postdoctoral thesis was published as *Makrosoziologie. Eine evolutionstheoretische Einführung* (Giesen 1980, "Macrosociology. An evolution-theoretical introduction"). The comparison between *Human Societies* and *Makrosoziologie*, written 25 years earlier, is instructive because they deal with similar topics yet are vastly different in terms of theoretical perspective and style. What separates both books is not only Giesen's subsequent cultural turn but also his mature intellectual self-confidence, which cannot be reduced to nor separated from his institutional position as a tenured German professor, who pursued his intellectual interests and passion projects relatively unencumbered by peer-reviews and third-party funding.

Makrosoziologie (Giesen 1980) opens with a discussion of Popper's "three worlds," distinguishing between the material world, the world of subjective consciousness, and the world of symbolic constructs and rationality. Even later, Giesen would draw parallels between "culture" and Popper's third world—albeit devoid of any intrinsic telos toward rationality. The book continues with a detailed theoretical discussion of the concepts of society and evolution, so that the reader is already half-way through before they reach his discussion of different historical types of society—which is markedly different from *Human Societies* with its short introduction. Furthermore, in *Makrosoziologie*, Giesen uses "the common threefold" distinction between "primitive, traditional, and modern societies" (probably inspired by the work of the German sociologist Friedrich Tenbruck), which he explicitly rejects in the introduction of *Human Societies* (p. 41). He already distinguishes between four aspects for each type of society, but his classification in *Makrosoziologie* is based on the Parsonian AGIL-scheme, distinguishing between economy, politics, social integration, and culture.

In his early discussion of "primitive societies," we encounter topics familiar from *Human Societies*, the importance of gossip and gift-exchange, family and charismatic leadership, rituals and totems, and many more. Even back then, Giesen was already interested in and familiar with a wide range of anthropological literature—and, importantly, with Durkheim's *Elementary Forms of Religious Life*, which at that time was not translated into German but is cited in its French original. Well into the twenty-first century, German sociologists—Giesen being one of the few exceptions—regarded Durkheim primarily as the author of the *Division of Labor*, *The Rules* and *Suicide*, while his later work was largely ignored. Giesen also exhibits some familiarity with anthropological classics, though more recent authors like Mary Douglas and Victor Turner, which later played an important role in his thinking about early as well as contemporary societies, received only a marginal treatment or were entirely absent. In contrast to the relatively egalitarian "primitive societies," "traditional societies" are marked by strong social hierarchies. They subsume very different types of societies as discussed in *Human Societies*, from early human settlements to full-fledged empires (and partially even territorial states). Modern societies, finally, are characterized by a shift from hierarchies to markets, which also entails the rationalization of knowledge as it becomes decoupled from social power and subject to critical scrutiny.

Giesen's Middle Period: The Turn Toward Cultural Sociology

In 1982, at the age of 34, Giesen became professor for macrosociology at the University of Giessen (and perhaps the youngest full professor in sociology in German history). His collaboration with Jeffrey Alexander, at that time based at UCLA and still committed to neofunctionalism, began in the first years at Giessen and culminated in a co-edited volume on *The Micro-Macro Link* (Alexander et al., eds., 1987). Despite their early encounter, both authors later moved independently toward a cultural sociology, after which their collaboration renewed and intensified. In his own contribution to the volume, Giesen is already moving toward a conception of culture as categorical and symbolic macro-structure (1987). At the end of the 1980s, Giesen became involved in the study of collective identity as leader of a research cluster on "national and cultural identity as a problem of European modernity" (1988–1991). It is during this period, that Giesen reworked his concept of culture and embarked on his cultural turn.

Of crucial importance is the year 1991, in which Giesen published *Die Entdinglichung des Sozialen. Eine evolutionstheoretische Perspektive auf die Postmoderne* ("The de-reification of the social. An evolution-theoretical perspective on postmodernity," Giesen 1991b)—his first monograph after a break of 11 years. While only available in German, the book was seconded by an English article titled "Code, Process and Situation in Cultural Selection" (Giesen 1991a), which offers a summary of the theoretical framework employed in *Entdinglichung* and a brief case study on German national identity, which was later developed into a separate monograph (Giesen 1998/1993). For the first time, Giesen distinguished between codes, processes, and situations which led him to a new understanding of culture and evolution vastly different from *Makrosoziologie*. Culture is conceptualized through codes, systems of signs and distinctions, which are "*low in energy* but *high in information*" (Giesen 1991a: 173). Cultural codes are applicable to an unlimited number of temporal processes, such as social practices, which are in turn responsible for the reproduction and variation of information. In contrast, situations are temporally and spatially defined and account for the selectivity of codes and processes, which must prove their pragmatic value in order to be reproduced in the future.

The implications of this new conception of cultural evolution are explored and elaborated in *Entdinglichung* (Giesen 1991b), which is a rather abstract and challenging work. Of crucial importance for cultural evolution is the gradual decoupling of code, process, and situation, which not only allows for different social practices to be enacted in the same situation, but also for different cultural codes being employed to interpret and evaluate the same social practice. It is precisely this process, which according to Giesen leads to the "de-reification of the social" and the postmodern condition of contemporary society. In doing so, Giesen self-consciously offers a modernist explanation of postmodernity.

The year 1991 marks an intellectual revolution and cultural turn in Giesen's thinking. The introduction of symbolic codes, replacing each other due to the selective pressures of action and its environments, allows—in principle—older codes (and practices) to survive and reproduce in specific kind of situations as evolutionary niches. This builds a bridge to *Human Societies*, which no longer operates with such a strict notion of cultural selection nor postulates a linear process like decoupling but embraces instead noncontemporaneity, the coexistence of new and old cultural patterns. In *Human Societies*, Giesen is no longer a modernist, looking for simple explanations, nor a postmodernist, who gave up on explanation entirely, but looking for a third way that allows for generalization and the recognition of patterns in human history.

Giesen's work on collective identity coincided with the renewed public interest in German identity in the wake of Germany's reunification. Thus, Giesen devoted his next monograph—later translated as *Intellectuals and the German Nation* (Giesen 1998/1993)—to the construction of German collective identity. For Giesen, the formation of the German nation happened under conditions analogous to Eisenstadt's historical studies of axial age civilizations (1982) as German intellectuals at the periphery started to criticize the political and economic centers in the name of an abstract national unity. This creative adaption of Eisenstadt's work was followed by a close collaboration between both authors that resulted in an article on "The Construction of Collective Identity" (Eisenstadt and Giesen 1995). Here both authors conceive of collective identity in terms of symbolic codes, distinguishing between primordial, civic, and cultural codes, which are based on presumed naturalness, familiarity, and the sacred, respectively (1995: 77–84).

A slightly different triad is employed by Giesen in his German monograph *Kollektive Identität* (1999), where he speaks of primordial, traditional, and universalistic codes. This resembles the aforementioned distinction between primitive, traditional, and modern societies, though universalistic codes already emerged in axial age civilizations. While these codes seem to imply a historical succession, from the primordial to the traditional to the universalistic, the historical analysis in the book reverses that order. Universalistic codes are discussed in the context of the French and German enlightenment, traditional codes as dominating the later German Empire, and primordial codes as informing modern antisemitism, first in France and then—with catastrophic consequences—in Germany. Primordial and traditional codes do not simply vanish when universalistic codes appear, but remain an option in the cultural repertoire of modern societies. Giesen's interest in collective identity and its often traumatic and triumphant roots leads us to the last stage of his intellectual biography, in which *Human Societies* was conceived.

Giesen's Mature Work: The Cultural Sociology of the Extraordinary and Ambivalence

The final period in Giesen's work is marked by his move to the University of Konstanz, where he served as full professor for macrosociology from 1999 until his retirement in 2013. The beginning of this period is also characterized by a renewed collaboration with Jeffrey C. Alexander, who also relocated at the same time to Yale University, starting with the co-authored volume *Cultural Trauma and Collective Identity* (Alexander et al. 2004). Alexander and his school became an important influence but also sparring partner for Giesen's mature thought. Giesen took inspiration from the turn toward narrative in the early days of the strong program and collaborated with Alexander on the development of sociological theories of performance (Alexander et al. 2006) and iconicity (Alexander et al. 2012). While he used to conceive culture primarily in terms of codes and classification, he increasingly understood also symbolic representation in the form of ritualization, mythization, and iconization as part of culture (Giesen 2010: 56–66). Thus, the modes of "representation" in *Human Societies*, which not only sets it apart from *Makrosoziologie* (1980) but also from much of his cultural-sociological work in the 1990s.

While Giesen was clearly inspired by Alexander's work, it also pushed his thinking into a different direction that resulted in a creative tension between his and Alexander's vision of cultural sociology. While both agreed on the importance of cultural codes, Giesen—under the influence of authors like Serres, Turner, Castoriadis, and perhaps Simmel—became increasingly interested in phenomena that defied symbolic binaries and cultural classification, phenomena characterized by in-betweenness, ambivalence, and ambiguity. This interest is clearly formulated in the introduction to *Human Societies* and is also a recurring motif in the following chapters:

> Some phenomena are, however, hard to fit into this natural grid of our regular reality. [...] These objects and persons that, according to a particular society, do not fit into its classificatory grid tell us sometimes more about its culture than those that neatly match its assumptions of what is regular, ordinary and real. In dealing with classification, we will, therefore, focus not only on the central distinctions established by a society—such as their definition of inside and outside or of heaven and earth, but we will also ask how societies imagine the exceptional individual and we will try to find out how they conceive of forces, people or objects that do not fit into the regular classificatory grid of a given culture. (p. 40)

At this stage of intellectual development, Giesen's interest in social and cultural orders gradually shifted toward the extraordinary, which he first conceived as an integral part of social life, then identified as the driving force behind the processes of classification and communication and, finally and increasingly, as the foundation of social reality itself.

Giesen's intellectual turn toward the extraordinary was anticipated by his interests in collective identity and cultural trauma, but finds its most elaborate expression in his magnum opus *Zwischenlagen* (2010), which—not yet translated into English and perhaps untranslatable—is accompanied by a brief summary of the argument alongside five empirical vignettes as the concluding contribution of the *Oxford Handbook in Cultural Sociology* (2012). According to Giesen (2012: 788), the analysis of symbolic binaries can't be the last word of cultural sociology, because there is "something that transcends the successful ordering and splitting of the world into neat binaries," including the "extraordinary space between those opposites," which is populated by "monsters," "victims," and "heroes," and the "unclassifiable remainder" of any attempt of classification. Following Douglas on dirt as "matter out of place," Giesen conceptualizes "garbage" as a "scandal for the cultural order of things" (2012: 794) and invites us to rethink "seduction" as a space between autonomy and heteronomy beyond rational action. According to Giesen, Eve's seduction of Adam is not just the original sin but marks the beginning of society and history as we know it.

The German monograph *Zwischenlagen* (Giesen 2010) contains more than a dozen essays that offer variations on the theme, often examining historical changes of classifications systems and the shifting meanings of entities that do not fit the grid. So-called monsters, once feared, became objects of curiosity and then later of pity. Even if we do not readily accept Giesen's controversial claim that the social order is grounded in the extraordinary (and not the other way around), we should appreciate his extraordinary cultural sociology as a challenge to and complement for the "ordinary" cultural sociology as it is usually practiced.

From Hunter-Gatherers to Contemporary Society—The Sociocultural Evolution of Humanity

In order to understand the specificity of *Human Societies* it is helpful to use the distinction between social theory and theory of society, which has become fairly common in German sociological theory in the last decades. According to Reckwitz (in Reckwitz and Rosa 2023: 12ff.), social theory addresses the question "What is the social?" and "From which perspectives can it be analyzed?" In contrast, a theory of society builds on social theory asking the questions: "What are the structural features of society and particularly of modern societies?" and "What are the concepts with which these societies can be investigated?" For Giesen, "the social" encompasses interactions, institutions, and societies at large, but his favored perspective is that of cultural sociology, highlighting processes of meaning-making and cultural structures. In terms of his theory of society, Giesen rejects the sharp separation between modern and premodern societies as well as the "common threefold" distinction between "primitive, traditional, and modern societies" (p. 41)—which he himself used in *Makrosoziologie*.

Instead, he proposes six ideal types of societies (or seven, if we count "Global Regimes" in the Epilogue), which build on top of each other. Orthogonal to this typology, Giesen distinguishes between the four modes of production, classification, circulation, and representation for each type of society. In contrast to the Parsonian distinction between economy, politics, social integration, and culture as subsystems, the modes are conceived in a more abstract way, allowing the comparison between phenomena often considered distinct. Circulation, for example, comprises diverse phenomena such as money, migration, and media. Giesen's choice of modes also reflects his cultural sociological perspective: He stresses specifically that production includes the manufacturing of goods with a primary symbolic function, while the modes of classification, circulation, and representation are almost cultural by design. Culture ceases to be a subsystem of society, confined to its box, but becomes co-extensive with society as such.

Arguably, production is the most materialist of Giesen's modes, describing society's adaptation to (and shaping of) its natural environment. It addresses the energy consumption and infrastructure of a society as well as its production of material and symbolic goods. Classification largely overlaps with the concept of codes in Giesen's earlier works, referring to the distinctions and boundaries drawn by society itself yet taken for granted. As already mentioned, the late Giesen was not only interested in the classificatory grid that societies use to impose order on themselves and the world, but particularly in phenomena that fall through the grid, that defy established distinctions or that are located at the boundaries—"betwixt and between" as Turner (1980) would have said. It is this emphasis on the extraordinariness of certain phenomena against the backdrop of orderly classification, which is characteristic of Giesen's later work, epitomized in his German magnum opus *Zwischenlagen* (2010). While classification refers to codes that divide society and the rest of the world into neat little boxes, circulation encompasses all processes that cross the boundaries between those boxes. Representation, finally, captures efforts that strive to present a coherent picture of the world. Throughout the book, Giesen tries to show that there is an elective affinity between the different modes central for a societal formation. This even seems to be true in contemporary society with old forms re-entering the world historical stage. The American political commentator Ezra Klein (2025) once described the beginning of the second Trump term as a return of "Court Politics," reminiscent of the absolutist monarchs that ruled European territorial states (p. 153f.), while Trump's economic policies and tariff plans strongly resemble mercantilism, which aimed to shield the state from foreign imports:

> Most important, however, was a new policy that aimed at closing the state borders for products from abroad by raising high customs. Instead of importing goods from abroad and thus shifting the profits to foreign producers the citizens should be encouraged to buy goods made in the homeland. In order to ensure the necessary supply of goods and services and to foster economic self reliance, new crafts and industries were encouraged and established. While as few goods as possible should be imported, exports were supported. Thus, for the first time in history economic policies were deliberately set and implanted. The sound principle of spending less money than having income was transferred from the private household to the state. The state was conceived of as an economic actor in a field of international competitors. This politics of *Mercantilism* seemed sound and supported the crafts and manufacturers at home, but it impeded international trade. (p. 155, emphasis in the original)

In order to understand contemporary society, we need to understand history, the social and cultural formations that dominated in the past. Yet, we should also not fall for the modernist trap of Bloch and brand seemingly antiquated forms of social organization and culture as pathological or backward. Apparently outdated symbolic codes and cultural practices can be better adapted to a situation than contemporary ones. Mercantilist policies are increasingly attractive in a world of rising geopolitical conflicts, which is why President Biden followed in many regards the protectionist tendencies of the first Trump term.

From Hunter-Gatherers to Sedentary Communities

The lectures on "Simple Societies" are far more elaborate and extensive than the chapters on the other ideal types discussed in the book. This is not by accident. The late Giesen shared with the late Durkheim the conviction that the social and cultural forms found in the earliest societies, especially hunter-gatherer societies, form the basis of contemporary modern society. This emphasis on evolutionary continuity, encapsulated in the dictum that *nothing simply disappears*, is absent from Giesen's early monograph *Makrosoziologie* (1980). There, Giesen discussed hunter-gatherer societies alongside horticultural societies under the rubric of "primitive societies" (1980: 130–149), already showcasing his command of and interest in anthropological literature. In Giesen's later work, anthropology played an even bigger role, through the works of Clifford Geertz and, even more importantly, Mary Douglas and Victor Turner. The direct comparison with *Makrosoziologie* shows that many of the topics discussed there—notably charismatic authority, gossip, and gift-exchange—return in *Human Societies*, yet in a vastly different theoretical context. In his earlier work, Giesen still conceptualized culture as a sort of superstructure built and maintained by religious specialists (e.g., shamans), which stands in contrast to his later understanding of culture as codes, a cultural infrastructure that forms the taken-for-granted background of human action.

Simple societies are introduced as subsistence economies, based on hunting and gathering, which produce no notable surpluses. This mode of production requires mobility and allows only for few personal possessions. Hunter-gatherers live in small relatively egalitarian communities held together by kinship bonds. Arguably, the third lecture, which introduces "some important distinctions fundamental to all forms of cultural classification and that can be found not only in simple bands of hunters and gatherers, but in all human societies" (p. 48), forms the core of the chapter and marks a clear departure from the *Makrosoziologie.* Starting point for Giesen is the spatial distinction between the inside and the outside, which also shapes the attitudes and behaviors of hunter-gatherers toward in-group members and strangers. Built on top of this spatial classification is the evaluative distinction between the pure and the impure, which largely but not entirely map on each other. While the inside is usually considered pure and the outside is often conceived as impure, Giesen also discusses symbolic pollution as a consequence of transgressing the boundary from the inside to the outside (and vice versa). Similarly, rituals of purification, including the cooking of raw food, allow for a safe passage between the outside and the inside. Giesen notes with Mary Douglas, who famously defined dirt as "matter out of place," that purity and impurity can also be the function of the proper or improper location of an object:

> Wet earth on the carpet is impure dirt, in the garden is regarded as fertile soil; saliva is no problem as long as it stays in our mouth, while, out of our body, it is treated as disgusting and appalling, food turns to dirt if it sticks on a dress etc. (p. 50).

Thus, the distinction between purity and impurity is not simply a symbolic binary structuring the world, but often a second-order classification that is applied to objects and people that do not fit the ordinary classificatory grid. For Giesen, the Durkheimian distinction between the sacred and the profane is closely related to purity and impurity—and builds in certain ways on it. In contrast to the profane, "the sacred cannot be used, consumed, transformed or traded" (p. 52). Sacred objects and persons do not fit the ordinary classification system but constitute a class of their own. The sacred has also a special relation to the community, representing its identity. Giesen further revisits Weber's concept of "charismatic authority," conceptualizing it as an embodiment of the Durkheimian sacred by extraordinary persons. As the most basic form of rule (and the only one in simple societies), "charisma" continues to play an important role in modern politics.

From there, Giesen moves on to circulation in simple societies, focusing on marriage and gift-exchange. The institution of marriage not only allows persons to cross the boundaries between families, but enforces circulation through the incest taboo. Gift-exchange, on the other hand, is no longer conceived as a primitive form of economic transaction (as in *Makrosoziologie*), but conceptualized as a ritual, where a liminal object crosses social boundaries. Often gifts are wrapped, which stresses their liminality and the ritualistic aspects of gift-giving. In contemporary society, the symbolism and practice of gift-giving continues to exist, from brides being wrapped in wedding dresses (p. 64f.) to political donations (p. 66f.). Representation in simple societies takes the form of relics and icons, but also ritual performances, which not only allow for transitions but also symbolize the identity of a group and construct its social reality.

The next chapter, as the title "The Transition to Sedentary Life" already suggests, describes a gradual process rather than a finished state. Technological innovations such as horticulture and herding allowed for more settled ways of life but also for semi-nomadic life-styles. The transition to sedentary life is not a one-way street. Giesen would not have had any issues with Graber's and Wengrow's contention that early societies were able to and—in some cases—did in fact turn away from small-scale horticulture and returned to foraging and hunting (2021). As an ideal type, sedentary societies generate a modest surplus and allow for more personal possessions—and a higher degree of inequality. Giesen discusses—among other things—the increasing importance of gender in social classification, communal feasts (such as the potlatch) as a form of circulation and as a sociable way to dispose surplus and, finally, trophy-taking, which lives on in contemporary sports trophies and touristic souvenirs.

From Conquest to Empire

Conquest societies and empires as societal formations become only possible with the advent of agriculture that is able to produce reliable surpluses. These surpluses allow for larger and more permanent settlements as well as an increasing specialization and division of labor. In conquest societies, the horizontal division of labor is complemented by an unprecedented degree of vertical stratification. In terms of classification, the social world is divided between the high and the low, the rulers and the ruled, the masters and the slaves, between mobile warriors and local peasants. Beyond social stratification, verticality becomes also important as a general principle of classification, of the ordering the world. The world was often imagined as a great chain of beings with the gods on top—literally in the case of the Greek gods, which were said to inhabit Mount Olympus, the highest mountain peak in Greece.

As conquest societies became empires, this verticality was superseded by the distinction between center and periphery. Empires were generally more inclusive than conquest societies and opened pathways for social mobility, allowing people to move from the periphery to the center—and vice versa. In conquest societies, invaders were often able to conquer huge swaths of territory and subjugate local populations, but these military conquests were often short-lived. In contrast, empires were made to last and required large-scale investments in infrastructure, a standing army, efficient taxation, and bureaucracy. The members of the new bureaucratic elite were often recruited on the basis of meritocratic principles, exemplified by the entrance tests for civil servants in the Chinese Empire. Notwithstanding the central role of the emperor, personal rule was largely replaced by the impersonal rule of law and bureaucracy. The all-mighty gods either retreated from this world or became themselves subject to its transcendent laws. For both, conquest societies and empires, military conquest is of crucial importance—yet in very different ways. In conquest societies, mobile warriors (often on horses) made a living by plundering and subjugating local populations, often turning them into slaves. Empires, however, are usually driven by their civilizing mission and quest for imperial (not territorial) expansion.

Arguably, conquest societies and empires are a somewhat unusual choice for ideal types of societies—Giesen himself did not use these concepts in his earlier writings on "traditional societies" (1980: 150–181). It was Eisenstadt's work on the axial age (1982) which allowed Giesen to see the elective affinity between imperial rule, impersonal law, efficient bureaucracy, and universalistic religion. In contrast to "empires," the origins of the concept of "conquest societies" are less clear. In *Human Societies*, Giesen cites Eisenstadt and Jaspers in his discussion of the axial age, but there is no comparable reference pertaining to conquest societies. We believe that an important source of inspiration for Giesen was *Kultursoziologie als Kulturgeschichte* ("cultural sociology as cultural history"), the magnum opus of Max Weber's brother Alfred Weber, which has never been translated to English and is largely forgotten, even in German sociology. Only in recent years, two shorter theoretical expositions to his cultural sociology, written shortly before and after World War I, have been translated into English (Weber 2015/1913, 2012/1920). *Kultursoziologie als Kulturgeschichte* (Weber 1997/1935) offers—not unlike Giesen's *Human Societies*—a cultural sociological reconstruction of human history, from "the primitives" of prehistory through the "high cultures" of ancient history to modern society. Yet here the similarity between both works, which use very different conceptions of culture, largely ends. A. Weber pairs a modernist conception of civilizational progress with a vitalist and holistic understanding of culture, while Giesen rejects the modernist narrative and conceptualizes culture in terms of classification systems and representations.

Yet it is in *Kultursoziologie als Kulturgeschichte*, where we find a precursor to Giesen's conquest societies, namely the riding master peoples that according to A. Weber were responsible for the discovery of freedom and the emergence of so-called secondary cultures (1997/1935: 91ff.). While there is no reference to A. Weber in the entire manuscript, we can find an appreciative discussion of *Kultursoziologie als Kulturgeschichte* in Giesen's *Zwischenlagen*. In an essay on axial ages and the demarcation of epochs, Giesen (2010: 276) not only mentions A. Weber as a precursor of Jasper and Eisenstadt, but also points out that Weber's conception of "secondary breakthroughs" (2010: 278) has advantages over the works of the latter, being able to account for Islam and the Reformation as axial breakthroughs long after the axial age. Giesen (2010: 279) proposes a structural understanding of axial ages, which is no longer bound to a specific time period, but defined by the role of intellectuals at the periphery, which criticize the political and economic center in the name of an abstract unity.

While Alfred Weber's *Kultursoziologie als Kulturgeschichte* has been largely forgotten and never translated into English, a summary of the argument pertaining to Giesen's conquest societies can be found in Jaspers' introduction to his *The Origin and Goal of Human History* (1953), where he discusses A. Weber's "riding master peoples"—Giesen's "mobile warriors"—as potential solution to the problem of the simultaneity of the axial age breakthroughs:

> In response to the question, why this simultaneity? only one methodologically arguable hypothesis has so far been advanced, that put forward by Alfred Weber. The penetration of the nations of charioteers and horsemen from Central Asia—which did, in fact, reach China, India and the West and introduced the horse to the ancient civilisations—had, so he argues, analogous consequences in all three regions. The men of these equestrian peoples came to experience, thanks to the horse, the limitless vastness of the world. They took over the ancient civilisations by conquest. In hazards and disasters they experienced the problematic character of existence, as master-peoples they developed an heroico-tragic consciousness that found expression in the epic. [...] Before, from Europe to China, there had been the ancient civilisations reaching back into the depths of the past and characterised variously as matriarchal, as civilisations of settled cattlebreeders, or simply as the population masses flourishing in closed self-sufficiency in the fertile regions of the belt of civilisation extending from China to Europe. History became a conflict between these two forces: the old, stable, unawakened matriarchal powers against the new, mobile, liberating tendencies of the equestrian peoples which were rising into consciousness. (Jaspers 1953: 16)

Jaspers' description is eerily similar to the conflict between "mobile warriors" and "immobile peasants" (p. 100), which is at the heart of Giesen's conception of conquest societies. Yet, Jaspers is concerned with the origins and simultaneity of the axial age breakthroughs. While he admits that "Weber's thesis possesses a singular power of illumination arising out of its simple, causal explanation based on the human character of the life of the horseman" (Jaspers 1953: 17), he ultimately rejects this explanation as insufficient. He points to historical counterexamples, such as the lack of tragic consciousness and the epic as a cultural form in the first millennium BCE in China or to the fact that "equestrian people" played no role in Palestine. Thus, the axial age revolution in ancient empires can't be a direct consequence of the discovery of freedom by the mobile warriors of conquest societies. Furthermore, one could add that the particularized freedom of the "master morality" in conquest societies—A. Weber draws heavily on Nietzsche in his reconstruction of history—stands in a fundamental opposition to the "slave morality" that characterizes the universalist religions of the axial age. Giesen resolves this contradiction by distinguishing between conquest societies and axial age empires, each of which gave rise to vastly different forms of social organization and cultural classification. Viewed from this perspective, the axial age ceases to be a solitaire but was preceded by a comparable transition from sedentary to conquest societies—and is succeeded by the transition from territorial states to modern nation states as a modern "axial age" (Giesen 1998/1993).

Last but not least we would like to note that Giesen in *Human Societies* weighs in on the debate on the causes of the axial age, offering a novel and intriguing explanation. For Jaspers, the philosopher, the emergence of the axial age had to remain a spiritual mystery. In contrast, the sociologist Eisenstadt identified social and political causes of the axial age. Others have recently pointed to economic causes such as "increased affluence" (Baumard et al. 2015). Giesen offers an alternative and decidedly cultural explanation for the axial age as a consequence of the invention of writing and the proliferation of texts:

> Writing and principled law fostered the idea of society as an abstract and invisible entity separated from the visible and tangible world of everyday life where people had bodies, faces and names. In distinction to oral communication, written texts allow for disembodied and decontextualized communication. (p. 128).

Thus, the invention of writing in conquest societies prepared the ground for the distinction between an immanent world of people and things and a transcendent world of principles. Later, Giesen discusses the timelessness of the text as a major difference between oral and written communication:

> The temporal sequence is replaced by the timeless coexistence of signs in a text and the text, devoid of time and location, becomes the object of thinking. This situation fosters second order reflections that search for timeless, abstract and translocal principles hidden in the texts. Writing about these principles imagines a world that is structured by an invisible eternal order

> behind the surface of everyday life. Thus the interpretation of written texts by intellectuals can give rise to the axial age divide between the profane and the transcendental realms that we have discussed above. (p.142)

Giesen ultimately suggests that the axial age divide is a direct consequence of writing and the proliferation and interpretation of texts. The discovery of transcendence in the axial age is no longer a mystery, as it was for Jaspers, but an effect of the order of signs in a text.

From Territorial States to Modern Nation States

Arguably, the historical periods Giesen was most knowledgeable about and interested in were early modernity, starting with the territorial states of the Italian renaissance, and Koselleck's "saddle time," the period between roughly the mid-eighteenth and mid-nineteenth century, where the modern nation state was born. Historically, European territorial states did not emerge out of empires in the sense that Giesen uses the term. As already Voltaire proclaimed, the Holy Roman Empire was "neither holy, nor Roman nor an Empire." The so-called emperor did not rule over the territory but was elected by his peers, feudal lords. Medieval Europe was populated by conquest societies with personalized rule and bonds of loyalty, though it was imperial in the sense that it was united by Catholic Christianity, an axial age religion with missionary zeal. Before Giesen, Marshall Sahlins and Elman Service described feudalism as a regression to an earlier type of society:

> Feudalism is a "stage" only in a specific sense, a step in the development of one line of civilization. The stage of general evolution achieved prior to the modern nation is best represented by such classical civilizations as the Roman, or by such oriental states as China, Sumer, and the Inca Empire. In the general perspective, feudalism is only a specific, backward form of this order of civilization, an underdeveloped form that happened to have greater evolutionary potential than the others and historically gave rise to a new level of achievement. (Sahlins and Service 1990: 32)

Giesen would have avoided the modernist conception of "stages" altogether (p. 42) and simply would have described it as return to an earlier type of society—although, of course, not completely. The heritage of the Roman Empire never disappeared entirely, not only Catholic Christianity but also Latin as *lingua franca* and the Roman law remained dominant on the European continent until the rise of territorial states in early modernity.

Giesen was well aware of the fact that territorial states did not emerge out of the meeting of expansionist empires—at least not on the European continent. Yet, he starts the chapter on territorial states with a conceptually driven narrative, which is reminiscent of Hegel's dialectics in the *Phenomenology of the Spirit*, arguing that the expansion of empires comes to an end when they meet, recognize each other, and consolidate their borders. The missionary force with which empires pushed outward starts to turn inward triggering the process of modernization: Monarchic rule became absolutist, tightening political control over the entire territory; the limited availability of resources through conquest required more effective taxation; rulers extended their influence over the economy with mercantilist trade policies; developments in science increased the mastery over nature and resulted in productivity increases.

Yet, despite "the turn inwards," European territorial states soon found new possibilities for conquest and exploitation outside of the European state system. The rest of the world was "discovered" by Europeans as a periphery from which resources could be extracted and onto which power could be projected. The formation of colonial empires illustrates the hybridity of societal types (p. 146): Modern colonial empires have a clearly demarcated center, a territorial or nation state with fixed boundaries and restrictive codes of inclusion, but project their power over a boundless periphery. In contrast to empires of old, modern colonial empires where not inclusive. While they spread Christianity to the heathens, they were not intent on turning conquered peoples into citizens (as the Romans did) and the social distance between the colonizers and the colonized was often stark (and comparable to conquest societies). Still, European colonial empires where driven by a missionary zeal to bring European civilization to the peoples of the world. In this common project they competed against each other, each trying to prove the worth of their civilization.

The creation of the modern nation state was a topic of preeminent importance for Giesen, who published from the 1990s onward extensively on collective identity and especially German identity. Giesen (1998/1993) argued that the nation state was a project largely driven by intellectual elites at the periphery, which opposed the traditional power centers and created a new sense of collective identity—and with it new forms of inclusion and exclusion. Thus, the shift from the territorial state to the nation state shares structural similarities with the classical notion of the axial age discussed in the chapter on empires. For Giesen, an important feature of modern society is "functional differentiation," a term used in the lecture on "classification" in modern nation states. Nowhere in the manuscript he mentions Niklas Luhmann, who coined and popularized the term. Instead of citing Luhmann, perhaps a rather obscure author in an international context, Giesen quotes Max Weber on value spheres and the autonomy of science (p. 178f.), grounding his argument about functional differentiation in the work of an undisputed sociological classic.

Global Regimes and Contemporary Society

Human Societies ends with a short postscript on "Global Regimes," which stands outside the lecture structure of the book. It does not offer systematic analysis of contemporary society in all four modes, rather some ideas and observations concerning what one could call the first truly "global society." According to Giesen, global regimes (and events) are a defining feature of contemporary human society. He points out that global networks and institutions have already existed in the past, but insists that globalization has reached a new quality in contemporary society. Like the Catholic Church in the past claimed to represent all Christians, there are today countless transnational organizations with universal outreach in different sectors of contemporary society, from political bodies like the UN representing all nations to the *International Sociological Association* representing all sociologists. This transnationalism does not fit into the classificatory grid of nations nor is it reducible to relations between nations.

Giesen further discusses the astounding acceleration of communication and circulation (of goods as well as people) in contemporary society. Without the unprecedented speed (and relative affordability) of the current transportation system, mass tourism as we know it today would be impossible. Fast communication and an emerging global public sphere not only bring the distant suffering of others to our attention, enabling global solidarity and the construction of a "collective identity of humankind" (p. 198), but also allow local events to trigger global conflicts. Last but not least, globalization has also a paradox effect on cultural differences in terms of classification, facilitating not only cultural homogenization but also cultural hybridity and differentiation through processes like creolization. It is easy to see how relatively recent phenomena like social media fit into this picture, not just by accelerating the circulation of texts, images, and videos but also by promoting cultural homogeneity and solidarity (e.g., though internet memes or icons of suffering) as well as differentiation and conflict (e.g., through the formation of echo chambers or by triggering global outrage over local transgressions).

Human Societies is an unfinished work, not just because its author was no longer able to prepare it for publication, but also because of the, likely intended, incompleteness and openness of its postscript, which exists in a state of in-betweenness—being simultaneously part of and outside the lecture script. Perhaps indicative of the openness of human history itself, the epilogue should be regarded an invitation for the reader to continue Giesen's lines of thought and fill in the gaps—especially for developments after 2006, the year in which he concluded the work on his manuscript.

Perhaps unsurprisingly, given his sustained focus on classification, circulation, and representation through the entire manuscript, Giesen doesn't address the current mode of production at all. Already in 2006, the global economy was not just characterized by the global circulation and consumption of locally produced goods but is already in their production dependent on global supply chains—whose importance and fragility became evident during the Covid-19 pandemic (caused by the rapid global circulation of a virus). Furthermore, like empires dependent on roads as transportation infrastructure, the globalized world of the twenty-first century depends on a digital infrastructure, even for its production of material goods and the provision of ordinary services. Additionally, an ever-increasing amount of goods are no longer consumed as material objects but exist in a purely digital form. Digital technologies have changed the mode as well as the economics of production: Digital goods can be replicated without additional costs, which not only increases the return on the initial investment for each additional unit sold, but also enables alternative business models such as streaming services. Last but not least, a discussion of the current mode of production should perhaps include what Nancy Fraser (2022) calls the "crises of reproduction" in contemporary (capitalist) society, for example global warming, which threatens to undermine the natural environment humans rely upon, or the upcoming global demographic crisis, which is still mainly framed as a national problem.

Yet, *nothing simply disappears*. Giesen is crystal-clear about the fact that global regimes do not replace the nation state; rather, contemporary society is characterized by the fusion of and tension between national and transnational (and other) elements. Arguably, this is a dynamic process rather than a linear progress. Since 2016, the year of the Brexit referendum and the first Trump election, the world has shifted toward a re-nationalization of the world economy and other societal sectors. Yet, it is also too early for a swan song mourning the end of globalization. The last years have clearly shown that globalization cannot be simple undone despite the efforts and fantasies of populist nationalists left and right.

Furthermore, even earliest forms of human culture and social interaction such as rituals—"embodied performances [...], produced and experienced bodily by actors in a shared situation and at a local site" (p. 79)—are alive and kicking in contemporary society. In a global and digital world, Giesen notes, "locality and bodily presence even get increasing significance and attention" (p. 199), which is not only a driver of global tourism but also a motivator for the in-person attendance of live concerts, sports events, and academic conferences. The forced isolation during the Covid-19 pandemic not only accelerated the trend toward digital communication and remote work, it also led to a deeper appreciation and valorization of face-to-face interactions and bodily co-presence. Despite the opportunities offered by digital communication (e.g., lectures on YouTube), "seemingly 'outdated' oral modes of communication" (p. 199)—such as the series of lectures on *Human Societies* by Giesen at Yale—will continue to exist in the future. Likewise, apparently "outdated" material objects such as analogue records not only survive but thrive in our digital age (Bartmanski and Woodward 2015).

In the spirit of Giesen, we would like to discuss two more recent examples of noncontemporaneity: the imperial aspects of the United States and Putin's Russia respectively. According to Giesen, empires are societal formations, which are shaped by the binary opposition between center and periphery with no clear territorial demarcation between them. Historically, empires have been less interested in tightening control over all their territory or in its expansion but were more concerned with the projection of power and influence from the center toward a periphery. Empires are not simply expansionist territorial states but operate according to a deterritorializing logic. Under the second term of Donald Trump, the United States became increasingly obsessed with territorial expansion (Greenland, Canada, Panama). Employing Giesen's concept of "empire," it would be wrong to characterize these ambitions for territorial expansion as imperial. Like the mercantilist trade policies and strict immigration policies favored by the Trump administration, their conception of territorial expansion is premised on clearly demarcated and strictly enforced borders. The United States under Trump follows the expansionist logic of territorial states and not the deterritorialized logic of an empire—and neither the exclusive logic of a modern nation state, whose borders are defined by the nation, which in turn is often defined by its (alleged) common descent.

In the manuscript, Giesen discusses the United States (alongside Switzerland) as a somewhat unusual nation state, not build on the idea of common descent but on civic participation: "It is no coincidence that the US cutting the ties between citizenship and descent were, for most of its history, not only open to immigration but depending on it" (p. 176). Yet, he briefly entertains the idea "that the US with its missionary zeal and its global reach out comes closer to the structure of empires than to the ones of nation states" (p. 176). Elaborating on this thought, we can conceptualize the United States as a global empire within a globalized world. The rise of the United States to the status of a global superpower was intertwined with the creation of the global order in the aftermath of the World War II and accelerated once globalization gained traction after the fall of the iron curtain in the 1990s. Only in the post-war period, English became the *lingua franca* in many sectors of global society and the dollar the global reserve currency.

The centers of the American Empire—and yes, it makes sense to use the plural here—do not necessarily coalesce in Washington, its national capital and political center. Instead, the United States exhibits multiple centers in accordance with the modern principle of functional differentiation. The global center of the entertainment industry is Los Angeles, the financial center of the world (and seat of the UN) is in New York, science and education are concentrated in prestigious institutions predominantly located in the Northeast of the United States, while the San Francisco Bay Area has morphed from a mecca of the counterculture to the center of the global high-tech industry. To the extent that these global centers are located within the United States but attract talent and capital from overseas, it makes sense to speak of a global American Empire. Like the empires of old, the American Empire largely deterritorialized. While the United States also functions as a nation state with its borders and collective identity, the American Empire depends on the deterritorialized flow of goods, capital, and persons, primarily from the periphery to its centers—but also the other way around, projecting influence and power across the globe—including military power due to more than 100 military bases located outside its territory. Under Trump, the United States has been actively moving away from its role as (a center of) a global empire, potentially degrading its future place in the world to a province among many others.

Another modern state, which shares many features of an empire is contemporary Russia, which—to the detriment of many Russian nationalists—is a multicultural society with a relative openness toward minorities and a disregard for territorial borders—not only since the full-scale invasion of the Ukraine. The Czech playwright and former president Vaclav Havel—arguing in 2005 for limits to the expansion of the European Union—once cited Russia as a negative example: "In the past Russia caused a great deal of suffering to many nations and ultimately to the entire world by, among other things, not knowing precisely where it began and where it ended." (2007: 308f.). Interestingly, these remarks have been echoed by the Russian president Vladimir Putin, who (only half-jokingly) proclaimed in a televised speech: "Russia's borders do not end anywhere" (BBC 2016). Eight years later and almost two years into Russia's full-scale invasion of Ukraine, the BBC correspondent Steve Roseberg (January 15, 2024) tweeted a photograph of a billboard in Moscow displaying Putin with the same slogan. The German policy analyst and columnist Ulrich Speck, who retweeted the post, commented on it as follows:

> "Russia's borders do not end anywhere": The very definition of empire.
> Empires have a center and peripheries.
> Nation-states have borders.
> Borders are at the center of the post-imperial order laid down in the UN charter, made of sovereign states equal in rights. (Speck 2024)

Giesen couldn't have said it better. The Russian war of aggression not only defies the global political order but also the very conception of a sovereign territorial or nation state this order was built upon. Instead, it follows the noncontemporaneous logic of an empire.[3]

[3] Contemporary Russia's disregard for territorial integrity is evidenced by its tolerance of 'zombie territories' such as Transnistria, Abkhazia or South Ossetia, which are de facto under Russian control but not incorporated. While Russia annexed Crimea in 2014, it did not do so with the territories it occupied in the Donbas. While Russia did annex these territories in 2022 after the full-scale invasion of Ukraine, it did so alongside large swaths of territory it never had conquered or controlled, not only in Donetsk and Luhansk but also in Kherson and Zaporizhzhia. Thus, their formal annexation by Russia should not be understood as a grab for territory but as a form of imperial trolling, mocking the principle of territorial integrity itself. Importantly, this deterritorialized logic also applies to peripheral territories within the Russian Federation. Chechnya is officially a republic within Russia, but it is de facto ruled by its own autocrat (loyal to Putin) with its own laws in open contradiction to those on a

Nothing simply disappears. Giesen's *Human Societies* offers a novel lens through which we can watch contemporary history unfold. History does not repeat itself. Yet, in weaving the fabric of history, humans draw on a historical repertoire of social and cultural forms producing thereby recurring patterns—often in uncanny detail, for example, when we recognize the Prigozhin revolt in the midst of Russia's war against Ukraine as an exemplification of the structural tension between political and military leaders in empires (p. 124).[4] All of the social and cultural forms that have risen and fallen in human history remain with us. Understanding humanity's past allows for a deeper understanding of our present times—and possibly the future. Not necessarily because history repeats itself, but because it never disappeared entirely in the first place.

national level. Similarly, when Ukraine invaded the Kursk region, an internationally recognized territory of the Russian Federation, in 2024, this might have been a successful publicity stunt for the Ukrainian war effort, but it failed to shift Russia's attention away from the frontline in the Donbass—arguably due to Russia's imperial mindset. From the perspective of the undisputed center in Moscow as "the third Rome," Kursk is as peripheral as the Donbass. For Russian elites, internationally recognized borders do not matter as much as Russia's sphere of influence. Belarus does not have to become part of the Russian Federation if it remains in Moscow's orbit. Ukraine could have done the same but refused to do so, motivated by the classificatory principle of the modern nation state, which constructed Ukraine as an independent nation, as one people endowed with sovereignty and the right of self-determination.

[4] During the first years of the full-scale invasion of the Ukraine, the Russian secretary of defense Sergei Shoigu rose to prominence, not only as a war criminal but also as a visible symbol for the inclusion of ethnic minorities into the Russian imperial elite. Later, he had a public spat with Yevgeny Prigozhin, head of the Wagner paramilitary group, who criticized the war-management of the ministry of defense. After Prigozhin failed to get Putin to act upon Shoigu, the charismatic warlord instigated a short-lived rebellion marching with his troops toward the imperial center in Moscow. One day later, Prigozhin gave up, after reaching an agreement with Putin. Exactly two months after the rebellion, Prigozhin died in a plane crash with other Wagner officials. This episode is not just a historical drama driven by human motivations and actions, not just a story of loyalty and betrayal, but also history shaped by recurring social and cultural patterns. The Prigozhin revolt exemplifies the baked-in conflict between military leaders at the periphery and political leaders in the center of empires discussed in *Human Societies*.

Editorial Policy (with Francis Le Maitre and Kim-Claude Meyer)

We, the editorial team, are all former students of Bernd (Giesen), who have been profoundly shaped by his thinking and would like to share this experience with others, especially those who never had a chance to meet and discuss with him personally. Among the things that Bernd left behind, was the PDF of the manuscript *Human Societies*, as it was prepared in 2006 and used for his Yale lecture series in 2008. The manuscript is in a remarkably polished state—though not necessarily in the shape Bernd would have given it before its publication. As editors, we felt obliged to respect the integrity of the text and tried to keep our interventions as minimal as possible. We corrected the obvious orthographic and grammatical errors (of which there are very few), harmonized the manuscript according to American English spelling (which is the dominant in the manuscript, though Bernd used British spelling for some words), and checked his quotes (which were not always reliable).

We also needed to adapt the structure of the manuscript to the requirements of the publishing house. The original manuscript was structured for its use as a lecture script, consisting of 25 lectures with an epilogue. As the book will be available online, with each chapter being separately downloadable, it made sense to bundle all four lectures devoted to one ideal type into one chapter, especially as some lectures are too short to stand on their own. This leaves Giesen's short introduction (as first lecture) and the even shorter epilogue as separate chapters. In the original manuscript, each lecture had its own list of references, which we decided to merge at the end of each chapter. It should be noted that most of the references are not cited in the text itself, but primarily intended as suggestions for further reading and as an (incomplete) acknowledgment of the sources that Giesen himself used. We kept the bibliography at the end of the manuscript, but added an index on names and subjects for the readers' convenience.

Furthermore, we decided on a new subtitle for the book while keeping the original (though rather nondescript) "Twenty-five lectures"—perhaps a homage to Alexander's *Twenty Lectures* (1987) on sociological theory—as the title of "Part II" containing the edited manuscript. In the end, we chose—inspired by the title of Giesen's habilitation thesis (1980)—"An Evolutionary Introduction to Cultural Sociology" as subtitle for the book as it reflects Giesen's unique perspective on the history of humanity. The title reminds of Gerhard and Jean Lenski's classic *Human Societies. An Introduction to Macrosociology* (1974), which Giesen already referenced in his *Makrosoziologie* (1980) and might also have inspired the title of his lecture series and manuscript—yet they approach the same subject from a very different perspective. Most importantly, the new title makes it clear that *Human Societies* was intended as both: An evolutionary account of human history as well as an introduction to cultural sociology.

No one simply disappears. Bernhard Giesen will remain with us, in our memories and in his writings. He lives on in his friends, colleagues, students, and readers.

References

Alexander, Jeffrey C. (1987): *Twenty Lectures. Sociological Theory Since World War II*. New York: Columbia University Press.

Alexander, C. Jeffrey (2021): "In memoriam Bernhard Giesen (20. Mai 1948—26. Dezember 2020)". In: *Soziologie* 50(2).

Alexander, Jeffrey C., Dominik Bartmanski & Bernhard Giesen (2012, ed.): *Iconic Power. Materiality and Meaning in Social Life*. New York; Houndmills: Palgrave.

Alexander, Jeffrey C., Ron Eyerman, Bernhard Giesen, Neil J. Smelser & Piotr Sztompka (2004): *Cultural Trauma and Collective Identity*. Berkeley: University of California Press.

Alexander, Jeffrey C., Bernhard Giesen & Jason L. Mast (2006, ed.): *Social Performance. Symbolic Action, Cultural Pragmatics, and Ritual*. Cambridge: Cambridge University Press.

Alexander, Jeffrey C., Bernhard Giesen, Neil Smelser & Richard Münch (1987, ed.): *The Micro-Macro Link*. Berkeley: University of California Press.

Bartmanski, Dominik & Ian Woodward (2015): *Vinyl. The Analogue Record in the Digital Age*. London: Bloomsbury.

Baumard, Nicolas, Alexandre Hyafil, Ian Morris & Pascal Boyer (2015): "Increased Affluence Explains the Emergence of Ascetic Wisdoms and Moralizing Religions". In: *Current Biology* 25(1), 10–15.

BBC (2016): "Russia's border doesn't end anywhere, Vladimir Putin says". BBC, November 24, 2016; https://www.bbc.com/news/world-europe-38093468

Bellah, Robert N. (2011): *Religion in Human Evolution From the Paleolithic to the Axial Age*. Cambridge; London: Harvard University Press.

Eisenstadt, Shmuel N. (1982): "The Axial Age. The Emergence of Transcendental Visions and the Rise of Clerics". In: *European Journal of Sociology / Archives Européennes de Sociologie* 23(2), 294–314.

Eisenstadt, Shmuel N. & Bernhard Giesen (1995): "The Construction of Collective Identity". In: *European Journal of Sociology / Archives Européennes de Sociologie / Europäisches Archiv für Soziologie* 36(1), 72–102.

Fraser, Nancy (2022): *Cannibal Capitalism. How Our System Is Devouring Democracy, Care and the Planet–and What We Can Do about It.* London; New York: Verso.

Giesen, Bernhard (1980): *Makrosoziologie. Eine evolutionstheoretische Einführung.* Hamburg: Hoffmann und Campe.

Giesen, Bernhard (1987): "Beyond Reductionism. Four Modells of Relating Micro and Macro Levels". In: Alexander, Jeffrey C., Bernhard Giesen, Neil Smelser & Richard Münch (eds.): *The Micro-macro Link.* Berkeley: University of California Press, 337–355.

Giesen, Bernhard (1991a): "Code, Process and Situation in Cultural Selection". In: *Cultural Dynamics* 4(2), 172–185.

Giesen, Bernhard (1991b): *Die Entdinglichung des Sozialen. Eine evolutionstheoretische Perspektive auf die Postmoderne.* Frankfurt/Main: Suhrkamp.

Giesen, Bernhard (1998/1993): *Intellectuals and the German Nation. Collective Identity in an Axial Age.* Cambridg; New York: Cambridge University Press.

Giesen, Bernhard (1999): *Kollektive Identität. Die Intellektuellen und die Nation 2.* Frankfurt/Main: Suhrkamp.

Giesen, Bernhard (2004): "Noncontemporaneity, Asynchronicity and Divided Memories". In: *Time Society* 13(1), 27–40.

Giesen, Bernhard (2010): *Zwischenlagen. Das Außerordentliche als Grund der sozialen Wirklichkeit.* Weilerswist: Velbrück.

Giesen, Bernhard (2012): "Inbetweeness and Ambivalence". In: Alexander, Jeffrey C., Ronald N. Jacobs & Philip Smith (eds.): *The Oxford Handbook of Cultural Sociology.* Oxford: Oxford University Press, 788–804.

Graeber, David & David Wengrow (2021): *The Dawn of Everything. A New History of Humanity.* London: Penguin.

Harari, Yuval Noah (2014): *Sapiens. A Brief History of Humankind.* London: Random House.

Havel, Václav (2007): *To the Castle and Back. Reflections on My Life as a Fairy-Tale Hero.* New York; Toronto: Alfred A. Knopf.

Jaspers, Karl (1953): *The Origin and Goal of History.* New Haven; London: Yale University Press.

Klein, Ezra (2025): "Trump 2.0 and the Return of 'Court Politics'". *The New York Times,* January 10, 2015; https://www.nytimes.com/2025/01/10/opinion/ezra-klein-podcast-erica-frantz.html

Lenski, Gerhard E. and Jean Lenski (1974): *Human Societies. An Introduction to Macrosociology.* New York: McGraw-Hill.

Reckwitz, Andreas & Hartmut Rosa (2023): *Late Modernity in Crisis: Why We Need a Theory of Society.* Cambridge; Medford: Polity.

Rosenberg, Steve (2024): "This morning an electronic billboard on my way to work is displaying this Putin quote: 'Russia's borders do not end anywhere'". Tweet on *X*, January 15, 2024; https://x.com/BBCSteveR/status/1746784252312891463

Sahlins, Marshall D. & Elman R. Service (1990): *Evolution and Culture*. Ann Arbor: University of Michigan Press.

Speck, Ulrich (2024): "'Russia's borders do not end anywhere': The very definition of empire." Tweet on *X*, January 15, 2024; https://x.com/ulrichspeck/status/1746805116978532659

Turner, Victor W. (1980): *The Ritual Process. Structure and Antistructure*. Cornell: Cornell University Press.

Weber, Alfred (1997/1935): *Kulturgeschichte als Kultursoziologie*. Marburg: Metropolis.

Weber, Alfred (2012/1920): "Fundamentals of Cultural Sociology: Social Process, Civilizational Process, and Cultural Movement (1920)". In: Loader, Colin (ed.), *Alfred Weber and the Crisis of Culture, 1890–1933*. New York: Palgrave Macmillan, 165–204.

Weber, Alfred (2015/1913): "The Sociological Concept of Culture". In: *Cultural Sociology* 9(2), 258–270.

HUMAN SOCIETIES: TWENTY-FIVE LECTURES BY BERNHARD GIESEN

Introduction

Bernhard Giesen

This book deals with human societies, with their institutions, structures, and symbolic media. In distinction to most other introductory accounts of the evolution of human societies, it takes a cultural perspective. While not denying the importance of material factors for structural or institutional change, its focus is on the world of symbolic interaction, social communication, and cultural interpretation. Culture is conceived of as the signs and stories, the rules and distinctions that allow us to interpret our world in a meaningful way, to overcome critical situations in interaction, and to answer questions in an acceptable way. Culture serves communication as language does speech, but it is not identical with language. Instead, it provides a "second layer" of understanding that operates behind the rules of grammar and the vocabulary of a language. Usually, we do not notice culture when, in our communication, it creates an understandable world. Only when misunderstandings cannot be repaired, when we encounter a lack of understanding, when we leave the familiar realm of our own culture and enter the territory of an unknown, unfamiliar culture, only then do we notice that our culture provides us with an indispensable web of meaning. Again, the similarity between culture and language may be instructive. As long as we speak the same language we do not notice the special rules of grammar that guide our speech. Only when we meet a person who obviously speaks a different language or makes a mistake we become aware of the particular rules of our own language.

B. Giesen et al., *Human Societies*, Cultural Sociology,
https://doi.org/10.1007/978-3-032-07697-7_2

We will present these cultural foundations of human society in a fourfold manner. First, we will focus on the modes by which societies adapt to their natural environments and produce material goods. We will call this function "production." Here, the focus is on the technologies of food production and the ways of harnessing energy from a given environment. Not all material goods, however, are produced in order to ensure the biological survival of the population and to adapt to the natural environment. The primary function of many material objects is carrying and displaying symbolic meaning: they mark social distinction and symbolize what is considered to be desirable in a particular culture. While not neglecting the problems of material adaptation we will focus on these symbolic meaning of objects in different cultures: palaces and temples, ornaments and tattoos, bodies of automobiles and sculptures. Older modes of production are never completely replaced and abolished by the rise of a new one. Instead, they continue to exist albeit in a marginalized and encapsulated way. Modern societies that are based on industrial production allow for the practice of hunting or gardening although they do not rely on it for food production.

The second perspective on culture concentrates on the most basic distinctions and boundaries that a given type of society establishes and takes for granted. These distinctions permit us to structure and map the world of objects, but also to fit fellow humans into a social grid. We call this mapping function "classification." Classification amounts to treating the phenomena of our world as if they would naturally fit into neatly separated boxes. This separation of "boxes" is the basic grid of our ordinary regular reality. Some phenomena are, however, hard to fit into this natural grid of our regular reality. Many of them escape our attention in ordinary contexts. If we cannot ignore them any more we, taking the point of a particular culture, consider them to be exceptional or scandalous. These objects and persons that, according to a particular society, do not fit into its classificatory grid tell us sometimes more about its culture than those that neatly match its assumptions of what is regular, ordinary, and real. In dealing with classification, we will, therefore, not only focus on the central distinctions established by a society—such as their definition of inside and outside or of heaven and earth—but also ask how societies imagine the exceptional individual and try to find out how they conceive of forces, people, or objects that do not fit into the regular classificatory grid of a given culture. Societies differ greatly in both respects.

The third perspective deals with the modes by which objects or persons shift their position in the social grid or in the map of the world constructed by classification. Exchange and appropriation, taxation and transportation, marriage and mobility, printing and the Internet are examples of this function that we call "circulation." Circulation allows for crossing the structural boundaries within societies and by this move to integrate what otherwise risks to fall apart. Every type of society disposes of certain typical modes of circulation that are added to the basic modes (marriage and gift), but these new modes of circulation do not abolish the existence of older ones. Taxation did not oust marriage and gift giving as Internet commerce did not replace local retail stores and television did not drive away printed newspaper.

Finally, in a fourth perspective on culture, we will outline the ways and media by which cultures represent the world. Material objects, images, written texts, speech, and performances are among these media or modes of signification that provide societies with the means to represent absent or invisible objects, persons, and collectivities in particular ways. We will call this function "representation." Representation allows for imagining the absent and the past in the present and thus to increase the range and scope of communication. Again, the invention or the emergence of new modes of representation does not replace completely the old ones: writing did not oust images and rituals as movies did not abolish theatrical performances. The "older" modes of representations continue to exist although sometimes in a marginalized position or in a contemporary disguise.

These modes of production, classification, circulation, and representation will be outlined with respect to six different types of societies: simple band societies, horticultural societies, conquest societies, axial age empires, territorial states, and modern nation states. Each of these types features special modes of production, classification, circulation, and representation that generate particular dynamics of solving problems while producing, as side effects, new ones. Thus no type of society can be seen as superior to another in absolute terms.

Although this typology is more detailed than the common threefold one (i.e., primitive, traditional, and modern societies) used in most evolutionary accounts of human society, it is, of course, too simple to mirror the manifold forms of human society in a detailed way. The evolutionary perspective used here is therefore more a presentation of ideal types than an accurate description of historical processes. Throughout the course of history, real societies have frequently blended these ideal types. For example, a society may have been in most respects a conquest society while also demonstrating some features typical of empires. There are transitions and there are, of course, exceptional cases that do not fit into our model. Nonetheless, it accounts fairly well for the variety represented by most archaic and historical societies. (As the great anthropologist Mary Douglas once noted: with respect to any theory that I put forward, there will be an ethnographer who maintains that "his" tribe is different).

In our presentation of different types of human societies we will focus only on those modes of production, classification, circulation, and representation that initially appear in any given type of society. As already mentioned above, the evolutionary "earlier" patterns do not disappear after a society has crossed the threshold to a model of greater complexity. All of them remain with us. Gardening and gifts, fear of pollution and family ties, trophy-taking, and the rule of terror—to mention just a few—remain phenomena of contemporary societies even if they have been translated into a modern context or if they have moved from a central position to a more marginal one. Real historical societies consist of a juxtaposition and blending of institutions and cultural patterns that originated at different evolutionary periods but the members of these societies are mostly unaware of this difference of origin. Not only do ancient patterns survive in modern societies but also can relatively modern techniques and institutions be adapted by societies that we commonly regard as premodern: today, Inuit fishermen can use motorboats and mobile phones, Maasai warriors can appeal to legal courts in order to claim their rights, etc. Thus, the evolutionary perspective of this book should NOT be mistaken as a neat historical sequence of evolutionary stages that succeed one another and replace the preceding patterns entirely. There is no neat separation between the so-called primitive culture and the modern mind—as many early anthropologists assumed.

Another caveat applies: our evolutionary perspective focuses on novelty and sometimes describes long-term changes as if they had occurred within only a few years. When we speak of the "Neolithic revolution," for example, we refer to a process that took at least centuries, if not more. In particular, it is misleading to assume that people "decided" to change the culture of a society, that they "opted" to turn from hunting to horticulture or "chose rationally" to replace feudalism by modern bureaucracy. Instead, the book attempts to present a repertoire of cultural patterns that has unfolded over the course of many thousands of years.

Simple Societies: The Basic Patterns of Social Life

Bernhard Giesen

Production: Hunting and Gathering

Simple societies must adapt to their natural environments. In distinction to more complex and technologically more advanced societies, their only source of energy is the human organism. Their tools and techniques are relatively simple and light to match their nomadic lifestyle. Spear, bow, and arrow enable hunters to kill animals at a distance, stone knives can cut meat and hides and carve wood, the command of fire paves the way for cooking, and weaving and pottery complete the range of simple technologies. Metal tools are unknown. Most items are manufactured from materials that can be found in the immediate surroundings and are sometimes left behind when the band moves on because transportation is more cumbersome than making new tools when they are needed. Every adult person knows how to manufacture these tools and is, therefore, much more independent from other persons than members of complex societies with a highly developed division of labor. When conflict and dissent occur between two groups or families, one of them can decide to leave the other and establish a new society without risking its members' survival.

These tools and, in particular, the use of fire set the societies of hunters and gatherers apart from primate societies, but, like these nonhuman societies, they do not themselves produce the food they need. They do not yet know how to grow crops or keep livestock as walking providers of meat.

B. Giesen et al., *Human Societies*, Cultural Sociology,
https://doi.org/10.1007/978-3-032-07697-7_3

Instead, they roam plains and forests in search of game, berries, insects, mushrooms, grain, and fruits. In this way, they compete with other species, in particular with other predators, for food. In order to sustain a given population through hunting and gathering, simple band societies require large territories and the size of their population is quite limited so as not to restrict their mobility. On an average basis, hunters need a territory of 10 to 20 square miles per person. It is almost impossible to provide enough food for thousands of people if no mode of fast transportation is available: food in the immediate environment is quickly exhausted, hunting expeditions must leave the camp for days in order to reach remote hunting grounds and, if the hunt was successful, the prey must be transported back to the camp. There are also limitations for the daily distance covered by walking or running: Most hunters do not run more than 25 miles a day—which is roughly the stretch of a marathon. All this in general limits the size of band societies to small groups of less than 100 persons who live by hunting and gathering around the camp and move camp as soon as food resources appear to be exhausted. If the population grows beyond certain limits, it must be cut back by practices that appear cruel to the modern mind: infants and old people who cannot move along and hence are a burden for the nomadic bands are killed. In addition fertility must be restricted: sexual continence during the nursing period is a widespread practice among hunters and gatherers.

This interdependency of technology, population, and environmental resources may be termed the ecological triangle. In a given environment, the available technology limits the density and size of the population. Hunting and gathering provide very low returns in caloric energy for a given amount of human energy invested into the environment. The grade of "thermodynamic efficiency" generated by this mode of production and adaptation is the lowest of all potential options. A larger population can only be sustained if new technologies exploit additional environmental resources. The ecological triangle holds true not only for hunters and gatherers, but also applies to the more complex societies that benefit from the invention of horticulture and the use of irrigation to the breeding of new species of rice and wheat.

The nomadic life of hunters and gatherers has manifold consequences. Obviously, it does not foster the construction of durable huts or houses. The materials for the construction of shelter must be found on site and are, for lack of animal transportation, usually left behind when the band moves on. Wealth is a burden in the literal sense. For similar reasons, storage of food is very limited. If hunters return with a larger animal to the camp, the prey is shared not only with kinship group but also with other members of the band. A feast is arranged and every participant eats as much as he or she can. The limited range of storage also impedes the production of more food than is required for the survival of the group. The economy is focused on subsistence instead of surplus production. Consequently, economic inequality is minimal among hunters and gatherers. Individual property rights are very limited and concern mostly personal dress, decorative items, and sometimes women. Survival is a matter of central concern, child mortality figures can be high, and starvation and diseases can seriously challenge the existence of these small societies.

It would be misleading, however, to assume that all simple societies were moved predominantly by the quest for food and the risk of starvation. In some of these simple societies, only relatively little time is spent for collecting food. These hunters and gatherers "work" rarely more than four to five hours a day and their quest for food is highly intermittent. In most cases, only two thirds of the total population produces food while the rest is not concerned with food production at all. Even in Bushman society that has to cope with an arid and austere environment, the work of one hunter can support four to five people. Faced by the relative lack of pressure to work, the favorite pastime in these societies seems to be chatting, joking, and playing. The anthropologist Marshall Sahlins has therefore called theirs the "original affluent society," thereby hinting at the cultural and social barriers that led simple societies to abstain from surplus production.

> What are the real handicaps of the hunting-gathering *praxis*? Not "low productivity of labour," if existing examples mean anything. But the economy is seriously afflicted by the *imminence of diminishing returns*. Beginning in subsistence there spreading from there to every sector an initial success seems only to develop the probability that further efforts will yield smaller benefits. This describes the typical curve of food-getting within a particular locale. A modest number of people usually sooner than later reduce the food resources within convenient range of camp. Thereafter, they may stay on only by absorbing an increase in real costs or a decline in real returns: rise in costs if the people choose to search farther and farther afield, decline in returns if they are satisfied to live on the shorter supplies or inferior foods in easier reach. The solution, of course, is to go somewhere else. Thus the first and decisive contingency of hunting-gathering: it requires movement to maintain production on advantageous terms.

> But this movement, more or less frequent in different circumstances, more or less distant, merely transposes to other spheres of production the same diminishing returns of which it is born. The manufacture of tools, clothing, utensils, or ornaments, however easily done, becomes senseless when these begin to be more of a burden than a comfort. Utility falls quickly at the margin of portability. The construction of substantial houses likewise becomes absurd if they must soon be abandoned. Hence the hunter's very ascetic conceptions of material welfare: an interest only in minimal equipment, if that; a valuation of smaller things over bigger; a disinterest in acquiring two or more of most goods; and the like. Ecological pressure assumes a rare form of concreteness when it has to be shouldered. If the gross product is trimmed down in comparison with other economies, it is not the hunter's productivity that is at fault, but his mobility.
>
> Sahlins, Marshall D. (1972): *Stone Age Economics.* Chicago: Aldine-Atherton, p. 33

As we have already mentioned, problems of storage and transportation obviously restrict such a society's abundance of food and tools. But it is not only the lack of technologies that enables nomads to abstain from producing more food and tools than necessary for subsistence or survival. Surplus production is also prevented by a culture that focuses on sharing in the present instead of individually caring for a distant future. The economic subject in these societies is not the individual, but the family. Similar to most family dinners in contemporary societies, food is consumed in disregard of the distinction between those who produced it and those who did not. The right of the producer to his or her product is overridden by the commonality of the meal.

Division of labor exists only with respect to gender—hunting is usually a male concern, while collecting edible plants, cooking, and pottery are mostly women's tasks. Since all adult persons command the basic techniques of food gathering and tool production, economic dependency between families is low.

The small size of the population allows each member of the simple society to know all other persons by face and name. There is no society beyond the visible world of the camp and the hunting expedition. Encounters with strangers are rare, as is warfare between different band societies. There is a strong ambivalence in coping with the presence of strangers: the encounter is intimidating because it disrupts the most fundamental distinction between inside and outside of the community. The stranger is therefore kept at a distance—either by running away or by attacking him. Strangers who do not speak the language of the nomads are sometimes considered to be nonhumans—either gods or animals.

Classification: Purity and Pollution, the Sacred, and the Profane

The ambivalent attitude toward strangers hints at another function that all societies—whether large or small—must fulfill: the shared cognitive mapping of the world. Communication within human societies can proceed only if we can assume that we all map the world in a largely similar way. Even if a closer look revealed that this assumption is, at least partially, illusionary, we cannot dispense with this illusion. Minor differences between our perspectives cannot dissolve our basic certitude about a shared perception of the world—just as minor misunderstandings are not capable of challenging our basic confidence in understanding each other.

This assumption of a common view of the world is supported and reinforced by a shared grid of classifications that allows us to refer to phenomena, occurrences, and encounters as cases or examples of a general type. Jointly turning something from an amorphous phenomenon into "a case of a certain kind" establishes order and instills confidence—the world loses its chaotic nature and is turned into a comprehensible structure. When we classify something, we reduce its infinitely manifold aspects to very few and disregard the others. We act as if it would be a clear case of a certain kind and ignore the manifold possibilities of labeling it in a different way.

If for some reason, however, this operation of classifying the world fails, a feeling of uneasiness arises, fear spreads, and panic may occur. This is the case when we face or encounter people who or things which resist or run counter to our common classificatory schemes. They are regarded as uncanny and we try to keep them at a distance so that we are able to disregard their existence. Strangers who cannot be assimilated and fitted into a familiar grid frequently trigger these feelings of uneasiness and fear. They hint at the ambivalence and arbitrariness of boundary construction. What seems to be a clear case could also, from a different perspective, be classified differently and shifted to a different conceptual box.

In normal everyday life, however, these doubts are disregarded. Taking a common- sense attitude, we classify phenomena according to their obvious attributes and confirm mutually and repetitively this classification.

Classification presupposes a set of boundaries that enable us to perceive distinctions and similarities. In the following lecture, we will briefly outline some important distinctions fundamental to all forms of cultural classification and that can be found not only in simple bands of hunters and gatherers, but in all human societies. Using the most elementary dimensions of mapping the world we will distinguish between spatial, evaluative, and temporal classifications. The first refers to objects, bodies, and their relative position in space, the second refers to issues of identity or to the distinction between right order and deviant phenomena, and the third refers to change and stability, creation and decay.

> Two passengers in a railway compartment. We know nothing about them, their origin or their destination. They have made themselves at home and have commandeered the little tables, coat-hooks and baggage-racks. Newspapers, coats and bags lie around on the empty seats. The door opens and two new travellers enter. Their arrival is not welcomed. A distinct reluctance to move up, to clear the free seats and let the newcomers share them is evident. The original passengers, even if they do not know one another, behave with a remarkable degree of solidarity. They display a united front against the new arrivals. The compartment has become their territory, and they regard each new arrival as an intruder. Their consciousness is that of natives claiming the whole space for themselves. This view cannot be rationally justified. It appears all the more deeply rooted.
>
> Nevertheless, matters rarely get to the point of open conflict. The passengers are subject to a system of rules that is not dependent on them. Their territorial instinct is curbed both by the institutional code of the railroad and by other unwritten norms of behaviour such as courtesy. So only looks are exchanged and formulaic apologies muttered through clenched teeth. The new passengers are tolerated. One gets used to them. Yet they remain, even to a diminishing degree, stigmatized.
>
> Enzensberger, Hans Magnus (1994): *Civil Wars: From L.A. to Bosnia.* New York: The New Press, p. 105

The most basic spatial distinction is set by the boundary between inside and outside. It is at the core of the very operation of classifying things and objects. Something belongs in a certain "box" and is similar to other items in this box or it does not belong in this box because it is different and, therefore, "outside." In classifying something, we assimilate the items within the box and contrast this similarity to the variation outside. There might be certain differences between these items inside the box, but by common decision we consider them to be unimportant.

The "spatial" boundary between inside and outside acquires a special salience when it is extended to a social group's insiders and outsiders. The insiders are considered to be similar, in a certain important respect at least, while the outsiders have no common feature apart from not belonging to the group. The outside world appears to lack the structure and order that we assume for the inside—it is the sphere of wilderness and unpredictability. We can construct this social boundary in very different ways and allow for certain possibilities for crossing the boundary while banning others, but we cannot conceive of a society that includes everybody and excludes nobody. Even the most universalistic and inclusive vision of humankind has to exclude some people, for example, those who commit crimes against humanity.

Moreover, the inside and outside distinction does not only apply to the boundaries of total societies. It also extends to the internal classification of the social world: we distinguish men from women, children from adult persons, families from other families, neighbors from strangers, leaders from their followers, and friends from enemies. Here, too, some boundaries can be crossed and others cannot. Strangers can become neighbors and friends can turn into enemies, whereas crossing the boundary of gender is, in most societies considered impossible, or at least embarrassing.

Any crossing of the boundary between outside and inside results in a moment of ambivalence and uncertainty in which the arbitrariness of social boundaries may be revealed. To counteract this ambivalence, the crossing of boundaries is controlled by special rituals of passage that, in a first step, ceremonially dissolve the old ties of belongingness, then suspend the person crossing the boundary in an extraordinary position of in-betweenness and, finally, include the person in the new group. When one person gives another a gift, we see a trace of this ritual: the gift is wrapped by the giver for the moment of passage and unwrapped by the receiver. Gift-giving requires not only this wrapping but also minimal ceremonial interaction. If at the end of December, a husband transfers $ 500 to his wife's account, we would not consider this transfer to be a proper Christmas gift unless he ceremonially declares this transfer to be a gift in face-to-face interaction with his wife.

Frequently, the spatial distinction between inside and outside is related to an evaluative distinction that appears in all cultures: the classification of phenomena as pure or impure. We can define the distinction between pure and impure from different points of view. The first perspective focuses on identity and similarity: pure are all those things that seem to have an affinity with the community and hence can be consumed by its members. Impure are, by contrast, those things that can be consumed only after undergoing a ritual transformation of purification or those items that cannot be consumed at all. Washing, dissecting, and cooking are among those rituals of transformation that allow us to consume and appropriate previously impure or raw objects. The distinction between pure and impure is not a matter of hygiene or medical health. If we reject something as disgusting or inedible, we may justify this rejection with medical arguments or personal taste, but our prime, though unconscious, motive is our learned cultural grammar of purity. In some cultures, pork is treated as impure, in others it is lusted after as a delicacy; in traditional western cooking, raw fish is rarely consumed, whereas in Japan a fish needs only to be gutted and aestheticized before it can be enjoyed. This aestheticization is a cultural substitute for the fundamental transformation of the raw into the cooked. The raw represents nature, impurity, the outside, whereas the cooked represents culture, purity, the inside. We can eat raw things without a ritual transformation only if they are not seen as relatively similar to us: apples can be eaten raw, the dead dog cannot. If an animal was called by a name and if we had communicated with it we mostly exclude it from consumption—not because we consider it to be raw or disgusting but because it appears to be related to us.

Sometimes, the same item passes the boundary between pure and impure within minutes through a change of location: the remains of a fish that we have just consumed are reclassified as disgusting waste as soon as they pass from our plates into the garbage can. This shifts our attention to a second perspective on purity and impurity. Here, impurity does not refer to the matter of objects, but to the location of the object. Pure are all those things that are at their culturally defined proper and correct location, impure are all those who are not—although the same matter may be considered as pure in another location. Wet earth on the carpet is impure dirt, in the garden it is regarded as fertile soil; saliva is no problem as long as it stays in our mouth, while, out of our body, it is treated as disgusting and appalling, food turns to dirt if it sticks on a dress, etc. Impure things that cannot immediately be removed to their proper location are classified as waste, dirt, or garbage. Waste and dirt, in distinction to other profane things, are useless, formless, and meaningless—they lead a scandalous existence not because of their material features but because they challenge our cultural classification. Hence, they must be removed and shielded from our eyes, ears, and noses; we can stand the presence of waste only when we are able to ignore it. Waste, dirt, and garbage have no internal order—everything has been thrown together, decay takes over, and we wrinkle our noses at the spreading bad odor. Under extraordinary circumstances, however, what is treated as dirt or waste under ordinary conditions can be converted into something that carries a redeeming magical power: blood of a martyr, human excrements of extraordinary persons, decaying pieces of human bodies can be used as protection against diseases, etc. This conversion of the axis of evaluation is due to the idea that evil can be prevented by another evil that is purified by the extraordinariness of its user and framed by a particular container that prevents contagion.

Like the boundary between inside and outside, the evaluative distinction between purity and impurity or pollution, too, applies not only to objects such as food but also to people. Some people are defined as pure, innocent, decent, civilized, and proper, while others are dirty, impure, treacherous, deceptive, and wild. "Pure" people are the right ones to marry and to have dinner with, while the presence of "impure" people can pollute the place and affect the purity of the community. Therefore, distance must be kept to impure people and a boundary must be established to protect the internal space of a community against pollution and poisoning. Here, the similarity between the inside-outside distinction and the pure-impure distinction becomes obvious. The latter is nothing more than adding an evaluation to the distinction between inside and outside. This transposition can engender outrageous consequences: racism, anti-Semitism, and processes of ethnic cleansing. But we can find this transposition also on less scandalous levels in our everyday life. When visiting a foreign country, many people are much more demanding with respect to cleanliness than they are at home. Foreigners are suspected of being less clean, their food is supposedly prepared in an unclean way, and their bathrooms are dirty, etc. At home we tend to be less sensitive—it is our kin's odor and unwashed bodies and the earth on the salad is from our garden. Lack of cleanliness is, at most, an exception or a mistake. Negligence and impurity do not occur within our family; they are only to be found among strangers.

> Now is the time to identify pollution. Granted that all spiritual powers are part of the social system. They express it and provide institutions for manipulating it. This means that the power in the universe is ultimately hitched to

> society, since so many changes of fortune are set off by persons in one kind of social position or another. But there are other dangers to be reckoned with, which persons may set off knowingly or unknowingly, which are not part of the psyche and which are not to be bought or learned by initiation and training. These are pollution powers which inhere in the structure of ideas itself and which punish a symbolic breaking of that which should be joined or joining of that which should be separate. It follows from this that pollution is a type of danger which is not likely to occur except where the lines of structure, cosmic or social, are clearly defined.
>
> A polluting person is always in the wrong. He has developed some wrong condition or simply crossed some line which should not have been crossed and this displacement unleashes danger for someone. Bringing pollution, unlike sorcery and witchcraft, is a capacity which men share with animals, for pollution is not always set off by humans. Pollution can be committed intentionally, but intention is irrelevant to its effect—it is more likely to happen inadvertently.
>
> Douglas, Mary (1966): *Purity and Danger. An Analysis of Concepts of Pollution and Taboo.* New York; Washington: Frederick A. Praeger, p. 113

Purity and pollution, the cooked and the raw, culture and nature blend the spatial distinction between inside and outside with an evaluative distinction. Another important evaluative mode of classification that has a close affinity to the distinction between inside and outside is based on the distinction between the sacred and the profane. Profane are those things, items, places, and times that can be used for other purposes that can be transformed and consumed, dissected and compared, and eaten and traded. Profane things have no value as such, but only with respect to higher purposes and in comparison to other profane things. They are not given personal names and we cannot communicate with them. They are ordinary everyday objects.

In contrast, the sacred cannot be used, consumed, transformed, or traded. It is valuable in and of itself and cannot be compared with anything else. The sacred is the basis of identity and the original source of action, motion, and creation. Encountering it inspires awe and demands respect. We can approach it or address it only by observing special ceremonial practices. While in many religions, the sacred is represented by gods and the places or things that have been touched by them, the sacred does not necessarily require a personal embodiment. In simple band societies, it is frequently associated with overwhelming forces like thunder, storms, and the sun, and in contemporary western societies it is represented by individual personhood, by human life, and human rights. But even in contemporary societies, the sacred is not confined to the sphere of abstract values. It exists not only in a disembodied mode, but resides in particular places and things and is celebrated at particular times: the Temple Mount in Jerusalem is such a sacred site for the Jewish community, as is the Dome of the Rocks for the Muslim community. This veneration of the same site by different religions has engendered serious conflicts which cannot be resolved by compromise and compensation. Any attempt to solve the conflict between Israel and the Arab world by offering a compensation in exchange for giving up the claim to this holy site would be rejected indignantly: the sacred cannot be substituted or traded. Only outsiders can consider what is sacred for the insider as an object of compromise and exchange. The insider fights to keep the sacred sites shielded from profane view and use and tries to liberate them from occupation by nonbelievers. In the same way as purity and impurity is a matter of location, the sacred is a matter of community. The community and its sacred center are not to be separated. If, by historical contingency and violence, both are separated the community has to struggle for their reunification, regaining the land of the ancestors or fighting for the Promised Land, praying at the place where the prophet, the messiah, the Buddha had their vision, etc.

The drive to separate the sacred from the profane has its counterpart in the attempt to be close to what a community considers sacred. The sacred has a permeating and contagious quality—whoever is close to it, touches it, or owns it participates in its redeeming aura. This is the reason why some people wear amulets, why others pay insane prices for a relic of a saint or an item previously owned by a famous personality, and why many people strive to own a piece of art, an expensive fashionable dress, or a cult car. They do not do so because it is cheap, useful, or functional, but, instead, because they expect the close proximity of the sacred to inspire and to invigorate them. The power of the sacred will be with them. This urge to own a relic is not only driven by the law of contagion but it is also fostered by the idea that the sacred is integral, that is, its parts suggest strongly the existence of the whole. While we can use profane things like screws or panels without constantly imagining their previous technical context we forget the sacred whole when we are close to one of its parts. Because all embodiments of the sacred hint at the integral whole the sacred establishes a web of meaning in the world and this web differs strongly from the one resulting from technical use.

What holds true for the individual person applies to whole societies as well. Sacred places and items such as temples and churches or the relics of a community's founding heroes must be housed at its center—any separation of a community from its sacred center provides a strong challenge for overcoming this undesirable state and, if necessary, for re-conquering the holy land. No impure nonbeliever may control and even touch the sacred site and sacred objects because the sacred must be separated from the profane even if this demands bloodshed. This separation lies at the core of construing identity and meaning. A world view that fails to draw this distinction—whatever its interpretation may be—is doomed to decay, chaos, and absurdity.

The sacred can also be embodied in special persons who are regarded as being endowed with extraordinary powers. For these persons Max Weber coined the term "charismatic."

> The term "charisma" will be applied to a certain quality of an individual personality by virtue of which he is considered extraordinary and treated as endowed with supernatural, superhuman, or at least specifically exceptional powers or qualities. These are such as are not accessible to the ordinary person, but are regarded as of divine origin or as exemplary, and on the basis of them the individual concerned is treated as a "leader." In primitive circumstances this peculiar kind of quality is thought of as resting on magical powers, whether of prophets, persons with a reputation for therapeutic or legal wisdom, leaders in the hunt, or heroes in war. How the quality in question would be ultimately judged from any ethical, aesthetic, or other such point of view is naturally entirely indifferent for purposes of definition. What is alone important is how the individual is actually regarded by those subject to charismatic authority, by his "followers" or "disciples." [...]
>
> It is recognition on the part of those subject to authority which is decisive for the validity of charisma. This recognition is freely given and guaranteed by what is held to be a proof, originally always a miracle, and consists in devotion to the corresponding revelation, hero worship, or absolute trust in the leader. But where charisma is genuine, it is not this which is the basis of the claim to legitimacy. This basis lies rather in the conception that it is the duty of those subject to charismatic authority to recognize its genuineness and to act accordingly. Psychologically this recognition is a matter of complete personal devotion to the possessor of the quality, arising out of enthusiasm, or of despair and hope.
>
> Weber, Max (1978): *Economy and Society. An Outline of Interpretive Sociology*. Berkeley: University of California Press, p. 241f.

Like other embodiments of the sacred, a person's charisma is not just a genuine quality belonging to him or her, but instead it is imputed to this person by the community of followers. A person cannot be charismatic without this community of followers who try to be close to the charismatic person and who expect extraordinary deeds from him or her. Both the expectation of extraordinary deeds and the attempt to be close to the charismatic person, however, also engender major risks for the charismatic relationship; nobody can continuously work miracles and no human being is able to conceal his or her human weaknesses completely if closely and permanently watched by others, even if these others desire to foster the charismatic person's aura. Therefore, the charismatic person must retreat occasionally from the public attention of these followers in order to conceal his or her ordinariness.

Personal charisma is the most elementary imagination of the exceptional individual and it provides the only source of leadership in simple band societies apart from the authority of parents and elders. The chief must prove his charisma by demonstrating extraordinary skill in hunting and wise decisions in conflict solving. Gifts made to his followers support this leadership. Thus, the charismatic leader does not exploit others; instead, it is he or she who gives away his or her wealth. The ethnologist Lévi-Strauss described the chief of the Nambikwaras as "the poorest man in the band":

> Although the chief does not seem to be in a privileged position as regards material belongings, he must have at his disposal a surplus of food, tools, weapons and ornaments which, however small, may nevertheless be of considerable value in view of the general poverty. When an individual, a family or the group as a whole feel a desire or a need for something, they turn to the chief. It follows that generosity is the main quality expected of a new chief. This is the note which is constantly struck, and the degree of consent is determined by its harmonious or discordant resonance. There can be no doubt that, in this respect, the chief's capacities are exploited to the utmost. Group chiefs were my best informants and, realizing the difficulties of their position, I was keen to reward them generously. But I rarely saw any of my gifts remain with them for more than a day or two. By the time I took leave of a group, after living with it for a few weeks, its members had become the proud possessors of axes, knives, beads, etc. Yet, as a general rule, the chief remained just as poor as he had been when I arrived. Everything he had been given (and this amounted to considerably more than the average presents made to each individual) had already been extorted from him. This

> collective greed often drives the chief into a kind of despair. When this happens, his refusal to give is more or less tantamount, in the primitive democracy of the Indians, to the demand for a vote of confidence in a modern parliament. When a chief gets to the point of saying: 'I am not giving anything more! I am not going to be generous any more! Let someone else be generous instead of me!' he must indeed be sure of his power, since his reign is going through the gravest possible kind of crisis.
>
> Lévi-Strauss, Claude (1976): *Tristes Tropiques*. Harmondsworth: Penguin Books, p. 408f.

The sacred is commonly split into two opposing forces: the sacred that redeems and creates identity and its counterpart, the demonic source of evil, danger, and decay. For hunters and gatherers, however, the sacred is simultaneously frightening and redeeming, horrifying and uplifting.

> It matters little whether they are conceived as distinct personalities or as diffused energies; under either form they fulfil the same function and affect the minds of the believers in the same way: the respect which they inspire is mixed with love and gratitude. The things and the persons which are normally connected with them participate in the same sentiments and the same character: these are holy things and persons. Such are the spots consecrated to the cult, the objects which serve in the regular rites, the priests, the ascetics, etc.—On the other hand, there are evil and impure powers, productive of disorders, causes of death and sickness, instigators of sacrilege. The only sentiments which men have for them are a fear into which horror generally enters. Such are the forces upon which and by which the sorcerer acts, those which arise from corpses or the menstrual blood, those freed by every profanation of sacred things, etc. The spirits of the dead and malign genii of every sort are their personified forms.
>
> Between these two categories of forces and beings, the contrast is as complete as possible and even goes into the most radical antagonism. The good and salutary powers repel to a distance these others which deny and contradict them. Therefore the former are forbidden to the latter: any contact between them is considered the worst of profanations. This is the typical form of those interdicts between sacred things of different species, the existence of which we have already pointed out. Women during menstruation, and especially at its beginning, are impure; so at this moment they are rigorously sequestered; men may have no relations with them. Bull-roarers and churinga never come near a dead man. A sacrilegious person is excluded from the society of the faithful; access to the cult is forbidden him. Thus the whole religious life gravitates about two contrary poles between which there

> is the same opposition as between the pure and the impure, the saint and the sacrilegious, the divine and the diabolic.
>
> But while these two aspects of the religious life oppose one another, there is a close kinship between them. In the first place, both have the same relation towards profane beings: these must abstain from all contact with impure things just as from the most holy things. The former are no less forbidden than the latter: they are withdrawn from circulation alike. This shows that they too are sacred. Of course the sentiments inspired by the two are not identical: respect is one thing, disgust and horror another. Yet, if the gestures are to be the same in both cases, the sentiments expressed must not differ in nature. And, in fact, there is a horror in religious respect, especially when it is very intense, while the fear inspired by malign powers is generally not without a certain reverential character. The shades by which these two attitudes are differentiated are even so slight sometimes that it is not always easy to say which state of mind the believers actually happen to be in. Among certain Semitic peoples, pork was forbidden, but it was not always known exactly whether this was because it was a pure or an impure thing and the same may be said of a very large number of alimentary interdictions.
>
> Durkheim, Émile (1965): *The Elementary Forms of the Religious Life*. New York: The Free Press, pp. 455–457

Like the sacred, the demonic, too, is contagious, but we respond to it in a contrary fashion. We are repelled by it and fear its sickening aura. Consequently, we try to keep it at a distance, to reveal its secret invasion into our community, to ban it with the power of the sacred, and to confine it or to kill whatever we see as its embodiments. Attraction and repulsion are two responses triggered by the same law of cultural contagion that constructs and reconstructs the distinction between the sacred and the profane. As universal as the distinction may be, its interpretation varies vastly from culture to culture. What is sacred for the insider is profane for the outsider, who, in turn, venerates a sacred item that appears to be utterly profane and banal for others.

> Religious man experiences two kinds of time—profane and sacred. The one is an evanescent duration, the other a "succession of eternities," periodically recoverable during the festivals that made up the sacred calendar. The liturgical time of the calendar flows in a closed circle; it is the cosmic time of the year, sanctified by the works of the gods. And since the most stupendous divine work was the creation of the world, commemoration of the cosmogony plays an important part in many religions. The New Year coincides with

> the first day of Creation. The year is the temporal dimension of the cosmos. "The world has passed" expresses that a year has run its course.
>
> At each New Year the cosmogony is reiterated, the world re-created, and to do this is also to create time—that is, to regenerate it by beginning it anew. This is why the cosmogonic myth serves as paradigmatic model for every creation or construction; it is even used as a ritual means of healing. By symbolically becoming contemporary with the Creation, one reintegrates the primordial plenitude. The sick man becomes well because he begins his life again with its sum of energy intact.
>
> The religious festival is the reactualization of a primordial event, of a sacred history in which the actors are the gods or semidivine beings. But sacred history is recounted in the myths. Hence the participants in the festival become contemporaries of the gods and the semidivine beings. They live in the primordial time that is sanctified by the presence and activity of the gods. The sacred calendar periodically regenerates time, because it makes it coincide with the *time of origin*, the strong, pure time. The religious experience of the festival—that is, participation in the sacred—enables man periodically to live in the presence of the gods. This is the reason for the fundamental importance of myths in all pre-Mosaic religions, for the myths narrate the *gesta* of the gods and these *gesta* constitute paradigmatic models for all human activities. In so far as he imitates his gods, religious man lives in the *time of origin*, the time of the myths. In other words, he emerges from profane duration to recover an unmoving time, eternity.
>
> Eliade, Mircea (1959): *The Sacred and the Profane. The Nature of Religion*. New York et al.: Harcourt, pp. 104–106

Cultural classification, however, does not only result from spatial or evaluative boundaries such as inside or outside or the sacred and the profane. We experience the world not only by mapping things and people, but also by putting experiences and events into a temporal order. The temporal distinction between before and after is as crucial for human experience as the spatial distinction between inside and outside or the evaluative distinction between the sacred and the profane. Earlier episodes and events set the stage for later ones and each new event appears to be new only by contrast and in distinction to a previous situation. This temporal and sequential structure also patterns our modes of communicating: we tell stories that present the phenomena of the world in a sequential order. We, in the present, remember the past and expect the future. All these stories have a beginning and an end and we can hardly grasp and imagine an infinite time before us and after us. These limits to our imagination are possibly related to the awareness of birth and death that set limits to our existence.

Practices such as narrating stories, remembering the past, or planning the future refer to time as a sequential order and classification according to the distinction between before and after. This most basic distinction is, however, superseded by a distinction between three realms or spheres of time that can be found in most cultures.

First, there is the time of normal history and profane change. Here, real events appear as what they are, the world has an enduring existence, everyday experiences occur, scarcity rules, and labor cannot be avoided. This is the time to which we refer when we talk about reality, experience, and normality and for many people it seems to be the only sphere of time that matters at all.

But there is also another time hidden and forgotten in real, everyday life. This is the time of dream and myth, in which everything seems to be possible, in which identities change and scarcity is of no importance, the sacred time in which the gods talk to humans, and the original paradise from which humans have been expelled but to which they relate in their dreams and mythical stories about the beginning of history. Most adult persons are able to separate clearly and strictly the time of profane reality from the sacred time of dreams. Children, however, frequently confuse these realms and, even for adult persons, drugs or dancing can produce an ecstasy in which the boundary between dream and reality is blurred. But this "time out of normal time" is not just a realm of individual dream and fantasy. It is also deeply rooted in collectively shared mythical narratives about the original paradise, the founding hero, the chosen people, and the Promised Land. These mythical narratives address issues that, for principled reasons, cannot be treated within the frame of profane and real time: how did history begin, what are its driving forces, and does it have an underlying aim, etc. Answers to these questions can inspire cultural identity and political visions and move even adult and mature persons to perform extraordinary actions.

Finally, there is the time of decay and dissolution, when the demons take over and the original chaos returns, when nothingness and the void of meaning reign. This reference to the apocalypse, to the end of the world, to Armageddon, etc. can be found in many cultures. In individual dreams, it appears as a deep-rooted fear of death and decay, while in collective myths, it is represented by narratives of the growing power of the forces of evil, of a final struggle between the embodiments of good and evil, and of the end of profane history.

To conclude this chapter, let us treat one of the most basic institutions of social structure: family and kinship. Although they are centered around the biological relationship between mother and child, family and kinship result from processes of cultural classification that define binding relationships between social positions and produce a sense of belonging. All kinship systems are based on three types of distinctions: the distinction between parents and children, the distinction between men and women, and the distinction between family members and outsiders. These basic distinctions can, however, be used to construct quite diverse kinship structures. Lineage and name can refer to the family of the father or to the family of the mother, the father can be closely attached to the family or he can be just an occasional visitor, the number of generations living together can vary, the attachment between uncles and nephews can be strong or weak, and the relationship to cousins can be strictly regulated or it can be disregarded, etc. Kinship allows for a complex classification that goes beyond the distinction between inside and outside or sacred and profane. In contrast to these fundamental boundaries, kinship generates the internal grid of simple societies and thus provides the backbone of their social order. The degree of solidarity and of obligation to other persons in these societies is molded and patterned primarily by kinship relations. In the small societies of hunters and gatherers, these relations are well known. If, however, the size of the population grows to include more than a thousand people, the definition of kinship relations can become a difficult task. In such cases, highly complex artificial kinship systems enable one to determine one's social obligations to another person. Even contemporary western societies retain traces of these artificial kinship systems. When we speak of "fatherland" or "patriotism" or when we use the concept of "nation" as a community of descent, we refer to kinship ties in a metaphorical way that presumes attachment, belonging, and obligations of solidarity.

The power of kinship as the foundation of social structure and community results from its stability and irreversibility. Children can never be turned into the parents of their parents, only by ignorance (as in the case of Oedipus) we can marry our own mother, and, at least in the days before modern surgery, changing gender was in all societies confined to an exceptional masquerade. We cannot rid ourselves of our family ties even if we would like to. Although the ties of kinship are socially constructed, they appear to be unchangeable and "natural," independent of our voluntary consent. Few institutions evidence the paradox of social reality in such a striking manner: kinship is a social construction but its constructedness must be concealed. Consequently, it is taken for granted and considered natural.

Circulation: Marriage and Gift

Human interaction is directed by the ties that bind, but the distinction between inside and outside is not the only one that pre-structures the field of possible actions. In addition, all human societies possess modes of circulating or communicating things, people, and messages between different groups and individuals. While the existence of communication is trivial to note with respect to language (we will return to this at the end of this chapter), it is less obvious with respect to people and things. But this very circulation of things and people across the boundaries of families is of crucial importance for simple societies.

> The small nomadic bands of the Nambikwara Indians of western Brazil are in constant fear of each other and avoid each other. But at the same time they desire contact, because it is the only way in which they are able to exchange, and thus to obtain products or articles that they lack. There is a link, a continuity, between hostile relations and the provision of reciprocal prestations. Exchanges are peacefully resolved wars, and wars are the result of unsuccessful transactions. This feature is clearly witnessed to by the fact that the transition from war to peace, or at least from hostility to cordiality, is accomplished by the intermediary of ritual gestures, a veritable 'reconciliation inspection'. The adversaries inspect each other and, with gestures which still hint of combat, examine the necklaces, earrings, bracelets and feathered ornaments of one another with admiring murmurs.
>
> And from being arrayed against each other they pass immediately to gifts; gifts are received, gifts are given, but silently, without bargaining, without any expression of satisfaction or complaint, and without any apparent connexion between what is offered and what is obtained. Thus it is a question of reciprocal gifts, and not of commercial transactions. But a supplementary stage may be reached. Two bands which have thus come to establish lasting cordiale relations can decide in a deliberate manner to join by instituting between the male members of the two respective bands the artificial kinship relationship of brothers-in-law. Given the marriage system of the Nambikwara, the immediate consequence of this innovation is that all the

> children of one group become the potential spouses of the children of the other group and vice versa. Thus a continuous transition exists from war to exchange, and from exchange to intermarriage, and the exchange of brides is merely the conclusion to an uninterrupted process of reciprocal gifts, which effects the transition from hostility to alliance, from anxiety to confidence, and from fear to friendship.
>
> Lévi-Strauss, Claude (1969): *The Elementary Structures of Kinship.* Boston: Beacon Press, p. 67f.

The core institution of circulating individual persons across the boundaries of kinship groups is marriage. Through marriage, a man or a woman leaves his or her original kinship group and becomes a member of a new one—that of his or her spouse. Thus, marriage provides a way of connecting two different families or kinship groups. Again, it is group membership that provides the ties that bind, but these ties are no longer based exclusively on descent, birth, and origin. Instead, two families which, originally, had no mutual obligations now are connected by a person who has a double membership to a family of descent and to the family of his or her spouse. The rule of exogamy, that is, the necessity of marrying someone outside one's own family, applies within all societies. With respect to the relations within a regular family, this requirement finds expression in the incest taboo.

The decisive function of the incest taboo is not—as our medical world view suggests—the avoidance of organic degeneration, but instead the production of communal ties beyond the family of descent. The circulation of men and women through marriage between family groups provides a second layer of social glue in simple societies.

Because of its extraordinary importance for social integration, the selection of a partner for marriage may not be based on such volatile fundaments as personal feelings of love. Romanic love, although well known in premodern societies, became the basis of marriage only in modern culture, which programmatically devalues ties of descent and requires that individuals determine their own identity and their destiny. Like other passions, love is beyond reason and comparison and is closer to temporary madness than to sound deliberation. As an extraordinary event, it cannot be commanded, expected, or produced—it just happens and, after some time, it fades away. Because of its extraordinariness, it disrupts the regular social order and disturbs the routinized flow of everyday life.

In simple societies (and even in more complex societies), marriage was far too important to be decided upon by two individuals who happened to feel a mutual erotic attraction. Indeed, we might even today consider it to be highly irrational to rely on a volatile feeling of exaltation such as romantic love for one of the most important and consequential decisions of one's life. Therefore, in premodern societies it was the parents or other mature persons who chose the partners with regard to political, economic, or social considerations. Simple societies sometimes had very complex rules concerning exogamy that prescribed the group from which the marriage partner was to be selected. If a person of marriageable age was not available in this group, no marriage could be arranged. Thus, in small societies, some people remained unmarried because all possible partners belonged to the wrong group.

This central passage from one group to another is commonly framed by a ceremonial ritual attended by members of both families and administered by a representative of the community. Bride and groom put on special robes or finery, a solemn ceremony must be performed, and a festive meal unites all participants, etc. Like other ritual passages, marriage, too, is usually performed in a three-step sequence: first, the person undergoing an identity change is symbolically disconnected from the previous group; in the second step, the ritual transformation occurs; and, finally, the person is reintegrated into the community and the successful passage is celebrated by all participants.

> The life of an individual in any society is a series of passages from one age to another and from one occupation to another. Wherever there are fine distinctions among age or occupational groups, progression from one group to the next is accompanied by special acts, like those which make up apprenticeship in our trades. Among semicivilized peoples such acts are enveloped in ceremonies, since to the semicivilized mind no act is entirely free of the sacred. In such societies every change in a person's life involves actions and reactions between sacred and profane-actions and reactions to be regulated and guarded so that society as a whole will suffer no discomfort or injury. Transitions from group to group and from one social situation to the next are looked on as implicit in the very fact of existence, so that a man's life comes to be made up of a succession of stages with similar ends and beginnings: birth, social puberty, marriage, fatherhood, advancement to a higher class, occupational specialization, and death. For every one of these events there are ceremonies whose essential purpose is to enable the individual to pass from one defined position to another which is equally well defined.

> Since the goal is the same, it follows of necessity that the ways of attaining it should be at least analogous, if not identical in detail (since in any case the individual involved has been modified by passing through several stages and traversing several boundaries). [...]
>
> I have tried to assemble here all the ceremonial patterns which accompany a passage from one situation to another or from one cosmic or social world to another. Because of the importance of these transitions, I think it legitimate to single out *rites of passage* as a special category, which under further analysis may be subdivided into *rites of separation*, *transition rites*, and *rites of incorporation*. These three subcategories are not developed to the same extent by all peoples or in every ceremonial pattern. Rites of separation are prominent in funeral ceremonies, rites of incorporation at marriages. Transition rites may play an important part, for instance, in pregnancy, betrothal, and initiation; or they may be reduced to a minimum in adoption, in the delivery of a second child, in remarriage, or in the passage from the second to the third age group.
>
> Van Gennep, Arnold (1960): *The Rites of Passage. A Classic Study of Cultural Celebrations*. Chicago: University of Chicago Press, pp. 2f., 10f.

The dissolution of the old ties can be symbolized by enclosure in a special room, by removing one's usual attire, by fasting, or—as in the Western ritual of marriage—by being escorted by one's father when processing to the altar. When somebody is disconnected from familiar social ties and undergoes this ritual transformation, for a moment at least, he or she is in an indeterminate and extraordinary situation. He or she is no longer member of his or her old group, but not yet a member of the new group, and thus is momentarily outside of the commonality and solidarity of kinship groups. In this time out of regular time, in the outlands of the community, we are alone, we encounter extraordinary powers, and we must face dangerous challenges in order to reach our destination.

Even in cases where this time of ritual transformation (as in marriage in Western societies) is reduced to a short moment, we are grateful for support and assistance—in the Western tradition, this is the task of the best man and the maid of honor. The extraordinary situation of in-betweenness, of being unclassifiable in accordance with the common social grid is usually shielded from the eyes of the community. Phenomena that escape or transcend the classificatory grid cause confusion and repulsion; if they cannot be ignored, they are regarded as monstrous or as garbage. Therefore, the moment of transition and transformation is frequently shielded from the common eye. Occasionally, this shielding is reduced to a wrapping or a special robe that conceals the body and face of the bride until the marriage is performed. We will return to this wrapping when we deal with the exchange of things.

However, this special bridal wrapping is necessary only in those cultures in which it is the bride who passes the boundary to the groom's family, takes on his name, and moves to his home. These cultures—and Western culture is one of them—are called *patrilinear* and *patrilocal* because they construct the lineage of descent by males and link the locality of the new couple to the husband's family. Here, as in other contexts, naming is of crucial importance. Beyond the grid of language, reality is indistinct and difficult to perceive.

Marriages are, of course, not the only rituals of passage. Baptisms, funerals, examinations, and investitures into office are other cases of such rituals of passage that transform one's social identity and position. Furthermore, circulation across the boundaries of kinship units is even in simple societies not limited to the passage of persons. Of similar importance is the transfer of things from one family to another. Precious and decorative items are passed from one family to another, sometimes in exchange for another gift. However, there is no way of precisely comparing the value of these directly exchanged gifts. If, for example, a person gives \$ 100 to another person at Christmas in order to receive in direct exchange another \$ 100 from this person, we would not consider this to be a proper exchange of Christmas gifts. Gifts differ strongly from barter or from market exchanges. Often, they are given even without the direct expectation of another gift in exchange or, at least, without the expectation of immediate return. This lack of any direct and accurate reciprocity brings us to the general meaning of gifts. The gift is meant to create a diffuse obligation between kinship groups or persons.

> *The obligation to give is the essence of the potlatch.* A chief must give potlatches for himself, his son, his son-in-law, or his daughter, and for his dead. He can only preserve his authority over his tribe and village, and even over his family, he can only maintain his rank among the chiefs—both nationally and internationally—if he can prove he is haunted and favoured both by the spirits and by good fortune, that he is possessed, and also possesses it. And he can only prove this good fortune by spending it and sharing it out, humiliating others by placing them 'in the shadow of his name'. Each Kwakiutl and Haïda noble has exactly the same idea of 'face' as has the Chinese man of letters or officer. It is said of one of the great mythical chiefs

> who gave no potlatch that he had a 'rotten face'. Here the expression is even more exact than in China. For in the American Northwest, to lose one's prestige is indeed to lose one's soul. It is in fact the 'face', the dancing mask, the right to incarnate a spirit, to wear a coat of arms, a totem, it is really the *persona*—that are all called into question in this way, and that are lost at the potlatch, at the game of gifts, just as they can be lost in war, or through a mistake in ritual. In all such societies one hastens to give. There is not one, single, special moment, even apart from the winter solemnities and gatherings, when one is not obliged to invite one's friends, to share with them the windfall gains of the hunt or food gathering, which come from the gods and the totems. There is not one single moment when you are not obliged to redistribute everything from a potlatch in which you have been the beneficiary; or when you are not obliged to acknowledge by gifts some service or other, whether performed by chiefs, vassals, or relatives: all this under pain of violating etiquette—at least for nobles—and of losing rank.
>
> Mauss, Marcel (1990): *The Gift. The Form and Reason for Exchange in Archaic Societies.* New York: W.W. Norton, p. 39

Whoever receives a gift knows that the giver can expect a diffuse reciprocity, at least friendly and kind behavior, gratitude and sometimes solidarity, and support in future times of crisis and danger. Thus, the meaning of the gift is not an economic, but a political one: like marriage, it provides a substitute for a central political authority that enforces rules and ensures peaceful relations between different kinship groups. The principle of reciprocity imposes an obligation so strong and inescapable that, occasionally, people try to find excuses for not being able to accept the gift (and the obligations attached to it). Someone who receives a gift and does not reciprocate at all is regarded as a cheater with whom no more contracts can be made. Even in modern societies we feel the strong obligation of diffuse reciprocation created by gifts. The exclusion of the cheater is deeply rooted and operates even if the expected reciprocation consists of illegal actions. Organized crime operates mainly on the basis of a principle of reciprocity: cheaters are killed. While organized crime appeals to reciprocity in a positive way, the legal institutions of modern society must relate to it in a negative one. In modern societies, gifts are forbidden as a direct means of influencing political office and jurisdiction. They are banned as corruption. If gifts are allowed at all, they must be laundered by an intermediary institution, such as a political party, that legally accepts donations for electoral campaigns. Moreover, these donations must be performed in a publicly transparent and legally controlled way. Similarly, buying electoral support by directly paying money to the voters is considered destructive to the democratic system—in contrast to issuing laws that improve the economic condition of a large group of voters. The second case represents a gift—it operates indirectly and results in a diffuse feeling of gratitude on the part of the voters.

Of course, giving gifts in order to gain political support is not peculiar to modern political systems. Political leadership was always based on the ability to provide one's followers with benefits. The feudal system in medieval Europe is a case in point. The seigneur, that is, the princely owner of the land, provided his vassals with parts if it and in return expected their military support when he put out a call to arms. Here, too, the land was a gift and the military support was a matter of noble obligation only indirectly linked to it. This stands in contrast to the Condottiere and mercenary soldiers of early modern times who were directly paid for their services (and consequently lacked diffuse loyalties to their clients).

Sometimes, the requirement to provide gifts to their followers even resulted in harsh conditions for leaders. If available resources were scarce, as in simple nomadic societies, the leader had to give away most of his possessions and was occasionally the poorest man in the community. Later on, the princely ruler had to exploit his subjects to fund the courtly splendor that, in turn, was intended to attract and impress aristocratic followers. With the advent of modern society, these gifts to the nobility were converted into regular payments, thereby losing much of their original binding power, Since it was now a regular and steady benefit the payment was turned from a generous gratuity into a right which the receiver could claim to.

In our society, gifts are commonly not just objects for profane use, but instead items of aesthetic and decorative value. We think the gift should convey a sense of the giver and match the special personality of the receiver. In contrast to exchanges on markets where no personal obligations apply and the rapid turnover of money and goods is not forbidden but required, a gift should be kept by the person who receives it. Selling it at the next opportunity is considered to be offensive to the person who gave the gift. This hints at the individual personal commitment symbolized in the exchange of gifts. The gift is a carrier of sympathetic feelings and personal commitment. It is closer to the realm of identity than to the profane sphere of matters and things. This may also be the reason why we wrap gifts before we transfer them. Wrapping and unwrapping is a central ritual part of the ceremonial gift exchange. Sometimes the wrapping seems to be more precious than the gift's content. The reason that we pay such great attention to wrapping is related to the symbolic nature of gift-giving. Just as persons undergoing a ritual of passage are redressed or wrapped for the moment of transition, the gift, too, must be kept from view when it is passed from one person or group to another. It is no longer the possession of the giver and not yet the possession of the receiver. In contrast to economic exchanges, the gift represents an extraordinary commitment and establishes communal ties between giver and receiver. It does not fit into the regular grid of possessions and belongings, but crosses the boundaries, and in doing so it, does not obey the precise correspondence of give and take as required by economic exchange. Instead, it is an exceptional and extraordinary commitment. In this respect, the gift is similar to the exceptional situation of the person who has left his or her original community in order to be integrated into a new one. The extraordinariness of the movement between two locations must be concealed from the eye of those who are firmly positioned in the social grid.

To conclude this chapter, we will turn to the paradigm case of communication, that is, to language. Talking to each other, too, establishes a communal bond between persons that grows with time. Most oral communication is not meant to provide information and to instruct another person, but, instead, it is intended to construct a shared definition of the situation and communal ties between the participants. Talking to each other consists mostly of confirming and reconfirming, rather than negating or questioning, the views expressed by the other speaker. We dwell extensively on topics that are presumably unanimously agreed upon and for which we do not risk demonstrating in divergent views. Talking about the weather is one of these ways of assuring consensus when starting a conversation—the weather is the same for all persons in a locality and we cannot change it.

> I think that, in discussing the function of Speech in mere sociabilities, we come to one of the bedrock aspects of man's nature in society. There is in all human beings the well-known tendency to congregate, to be together, to enjoy each other's company. Many instincts and innate trends, such as fear or pugnacity, all the types of social sentiments such as ambition, vanity, passion for power and wealth, are dependent upon and associated with the fundamental tendency which makes the mere presence of others a necessity for man. Now speech is the intimate correlate of this tendency, for, to a natural man, another man's silence is not a reassuring factor, but, on the

contrary, something alarming and dangerous. The stranger who cannot speak the language is to all savage tribesmen a natural enemy. To the primitive mind, whether among savages or our own uneducated classes, taciturnity means not only unfriendliness but directly a bad character. This no doubt varies greatly with the national character but remains true as a general rule. The breaking of silence, the communion of words is the first act to establish links of fellowship, which is consummated only by the breaking of bread and the communion of food. The modern English expression, 'Nice day to-day' or the Melanesian phrase, 'Whence comest thou?' are needed to get over the strange and unpleasant tension which men feel when facing each other in silence.

After the first formula, there comes a flow of language, purposeless expressions of preference or aversion, accounts of irrelevant happenings, comments on what is perfectly obvious. Such gossip, as found in primitive societies, differs only a little from our own. Always the same emphasis of affirmation and consent, mixed perhaps with an incidental disagreement which creates the bonds of antipathy. Or personal accounts of the speaker's views and life history, to which the hearer listens under some restraint and with slightly veiled impatience, waiting till his own turn arrives to speak. For in this use of speech the bonds created between hearer and speaker are not quite symmetrical, the man linguistically active receiving the greater share of social pleasure and self-enhancement. But though the hearing given to such utterances is as a rule not as intense as the speaker's own share, it is quite essential for his pleasure, and the reciprocity is established by the change of roles. There can be no doubt that we have here a new type of linguistic use—*phatic communion* I am tempted to call it, actuated by the demon of terminological invention—a type of speech in which ties of union are created by a mere exchange of words. Let us look at it from the special point of view with which we are here concerned; let us ask what light it throws on the function or nature of language. Are words in phatic communion used primarily to convey meaning, the meaning which is symbolically theirs? Certainly not! They fulfil a social function and that is their principal aim, but they are neither the result of intellectual reflection, nor do they necessarily arouse reflection in the listener. Once again we may say that language does not function here as a means of transmission of thought. But can we regard it as a mode of action? And in what relation does it stand to our crucial conception of context of situation? It is obvious that the outer situation does not enter directly into the technique of speaking. But what can be considered as *situation* when a number of people aimlessly gossip together? It consists in just this atmosphere of sociability and in the fact of the personal communion of these people. But this is in fact achieved by speech, and the situation in all such cases is created by the exchange of words, by the specific

> feelings which form convivial gregariousness, by the give and take of utterances which make up ordinary gossip. The whole situation consists in what happens linguistically. Each utterance is an act serving the direct aim of binding hearer to speaker by a tie of some social sentiment or other. Once more language appears to us in this function not as an instrument of reflection but as a mode of action.
>
> Malinowski, Bronislaw (1989): "The Problem of Meaning in Primitive Languages". In: Ogden, C. K. and I. A. Richards (eds.): *The Meaning of Meaning. A Study of the Influence of Language upon Thought and of the Science of Symbolism.* San Diego et. al.: Harcourt Brace Jovanovich Publishers, 296–336, p. 314f.

People in simple societies take much more time than we do for talking, chatting, and joking with each other. But even in contemporary societies, we can observe the construction of communal ties through talk across social or territorial boundaries: neighbors talk when they meet each other at the borders of their territories and must find an excuse if they have no time for engaging in lengthy conversations. The same holds true for people who meet in a public place and know each other from previous encounters. Silence, in contrast, is almost universally regarded as a sign of social distance, even of hostility. The only exemptions from this rule are people whose familiarity with each other is obvious because they know each other or have lived together for a long time. All this demonstrates the importance of oral communication for the construction of communality across the boundaries of family groups. This holds true even for oral teasing or joking. When we are joking or mocking at each other we alleviate and express the strain or aggressive energy that is an inevitable side effect of interaction but we do so in an unreal mode of communication. Because we are not serious we can communicate something that, in a serious mode, would be seen as hostile and offensive. A joke or mockery at the expense of another person allows and even presses this person to hit back in a joking way which, on its turn, engenders follow up jokes, etc. Thus joking circulates the right to be aggressive within a group.

> The joking relationship is a peculiar combination of friendliness and antagonism. The behaviour is such that in any other social context it would express and arouse hostility; but it is not meant seriously and must not be taken seriously. There is pretence of hostility and a real friendliness. To put it in another way, the relationship is one of permitted disrespect. Thus any complete theory of it must be part of, or consistent with, a theory of the

place of respect in social relations and in social life generally. But this is a very wide and very important sociological problem; for it is evident that the whole maintenance of a social order depends upon the appropriate kind and degree of respect being shown towards certain persons, things and ideas or symbols.

Examples of joking relationships between relatives by marriage are very commonly found in Africa and in other parts of the world. Thus Mademoiselle Paulme records that among the Dogon a man stands in a joking relationship to his wife's sisters and their daughters. Frequently the relationship holds between a man and both the brothers and sisters of his wife. But in some instances there is a distinction whereby a man is on joking terms with his wife's younger brothers and sisters but not with those who are older than she is. This joking with the wife's brothers and sisters is usually associated with a custom requiring extreme respect, often partial or complete avoidance, between a son-in-law and his wife's parents.

The kind of structural situation in which the associated customs of joking and avoidance are found may be described as follows. A marriage involves a readjustment of the social structure whereby the woman's relations with her family are greatly modified and she enters into a new and very close relation with her husband. The latter is at the same time brought into a special relation with his wife's family, to which, however, he is an outsider. For the sake of brevity, though at the risk of over-simplification, we will consider only the husband's relation to his wife's family. The relation can be described as involving both attachment and separation, both social conjunction and social disjunction, if I may use the terms. The man has his own definite position in the social structure, determined for him by his birth into a certain family, lineage or clan. The great body of his rights and duties and the interests and activities that he shares with others are the result of his position. Before the marriage his wife's family are outsiders for him as he is an outsider for them. This constitutes a social disjunction which is not destroyed by the marriage. The social conjunction results from the continuance, though in altered form, of the wife's relation to her family, their continued interest in her and in her children. If the wife were really bought and paid for, as ignorant persons say that she is in Africa, there would be no place for any permanent close relation of a man with his wife's family. But though slaves can be bought, wives cannot.

Social disjunction implies divergence of interests and therefore the possibility of conflict and hostility, while conjunction requires the avoidance of strife. How can a relation which combines the two be given a stable, ordered form? There are two ways of doing this. One is to maintain between two persons so related an extreme mutual respect and a limitation of direct personal contact. This is exhibited in the very formal relations that are, in so

> many societies, characteristic of the behaviour of a son-in-law on the one side and his wife's father and mother on the other. In its most extreme form there is complete avoidance of any social contact between a man and his mother-in-law.
>
> This avoidance must not be mistaken for a sign of hostility. One does, of course, if one is wise, avoid having too much to do with one's enemies, but that is quite a different matter. I once asked an Australian native why he had to avoid his mother-in-law, and his reply was, 'Because she is my best friend in the world; she has given me my wife'. The mutual respect between son-in-law and parents-in-law is a mode of friendship. It prevents conflict that might arise through divergence of interest.
>
> The alternative to this relation of extreme mutual respect and restraint is the joking relationship, one, that is, of mutual disrespect and licence. Any serious hostility is prevented by the playful antagonism of teasing, and this in its regular repetition is a constant expression or reminder of that social disjunction which is one of the essential components of the relation, while the social conjunction is maintained by the friendliness that takes no offence at insult.
>
> Radcliffe-Brown, A. R. (1979): *Structure and Function in Primitive Society. Essays and Addresses.* London; Henley: Routledge & Kegan Paul, p. 91f.

The relatively rare cases in which oral communication is undertaken to transfer information from one person to another do not, however, follow the logic of circulation embodied in marriage or gift-giving. In contrast to things or persons who pass a boundary and are afterward no longer located in their original position, information is not lost for those who communicate it to others. The others may now possess the information, but we still have it, as well. Information is *replicated* instead of being *circulated.*

Oral communication presupposes that there are certain hints through which the next speaker perceives that the preceding speaker will not continue to talk and that the next speaker's turn has started. In most cases, these hints consist of a lowering of the voice. Without these turn-taking rules, the sequential order of conversation would be hard to establish, many speakers would talk simultaneously, and the resulting chaos would seriously impair understanding. Hence the turn-taking rules are at the core of oral communication.

Representation: Relics, Emblematic Icons, and Ritual Performances

Culture assumes that perceived phenomena are not just what they appear to be, but instead that they are carriers of meaning. Cultural items transcend immediate perception and represent something else. A pattern painted on a body is not just meaningless dirt, rather it hints at membership to a special social group; a piece of wood is not just a piece of wood, rather it represents the deity or the ancestors; a colorful dress is not just a useful protection against chilly weather, but a sign of leadership; a joint bodily movement is not just coincidental, but a ceremonial dance to attract rain; and a flag is not merely a piece of cloth, etc. Culture consists of these acts and items of symbolic representation and of the understanding of their meaning. Belonging to a cultural community means being able to act on and to understand these symbols, the meaning of which remains hidden to the outsider. Translation of this meaning to outsiders is, of course, possible, but it takes time, patience, and practice to become a recognized member of a cultural community and to act within the cultural web of meaning. The reason for this slowness is a principled one. Like the meaning of words in a spoken language, the cultural meaning of things and actions, too, can never be completely grasped by precise rules; there are exceptions from these rules, variations, and modulations of their application. If we tried to find rules for these exceptions and variations, we would discover further exceptions to and variations on these rules. Therefore, just as reading a dictionary is hardly a substitute for speaking a language, reading a book about a culture can hardly replace participating in its practice.

In the following lecture, we will not deal with the content of representations, but with the basic media of representation as they can be found not only in simple bands of hunters and gatherers but in all human societies. We will distinguish between relics, icons, and ritual performances as the three basic modes of symbolic representation.

Relics are parts or remainders of an absent whole. This original whole is represented by the material continuity of elements that were once a part of it. The absent, past, and decayed whole is represented by a material object that has been removed from its original context or continues to exist even when its original source is no longer present. Like a trace or a signal, the relic, too, is a product of natural sources, but, unlike the trace, it shares its own materiality with the original whole. Because of this natural causation and material identity, the relics seem to establish a particularly strong tie of representation. The represented item is undeniably present, although only as a part of the original whole.

Relics are important items within the restricted range of possessions in nomadic societies. Amulets and talismans made out of bones of totem animals protect their bearers from evil spirits and symbolize the identity of the carrier. Again, it is the principle of contagion that transfers the spiritual power of the amulet to those who are close to it or own it. Relics have an aura. They are communicative objects that stimulate emotions and inspire actions.

But, of course, relics are found not only in simple societies dominated by magical thinking. Representation by material relics is a powerful moving force in all societies. The bones of ancestors or saints venerated in special enclosures, the personal belongings of famous persons exhibited in museums, antiquities bought by collectors, and trophies and souvenirs proudly displayed at home, as different as they seem to be, all derive their attraction and value from the same cultural logic of relics. The relic is frequently the carrier of charisma. Its presence fuses the locality with sacrality, its possession inspires the owner, it attracts members of the community, and it can become the object of conflict. Its charismatic power results from its capacity to represent the absent sacred in material form. Hence, the issue of originality lies at the core of the communication about relics. An original fragment or torso of a statue is much more valuable than the complete and perfect copy, a souvenir must be bought on the site of the vacation and not at home, and a trophy must testify to the effort of its owner in acquiring it and loses its charismatic radiation if no effort was involved, etc. The value of the relic is drawn from material presence and the transfer of the sacred to its material embodiment.

Emblematic icons such as the cross, the national flag, a heraldic sign or totem are not linked to their reference by material descent, but by formal homology. They affirm social ties that, by their very nature, are invisible and thus must be represented by visible signs. These signs are relatively simple and standardized and hence easy to recognize. Everyone knows their meaning and can respond to their presentation with appropriate action. Icons have a relatively stable meaning, but this meaning is not very complex—it is centered on the affirmation of social ties and excludes the representation of temporal change, process, or negation. The most important emblematic icons in simple societies are totem signs. Totems represent groups by relating them to particular animals or other phenomena of nature that are sacred for the members of this group. Consequently, they cannot be eaten or killed by group members, whereas they remain profane objects for outsiders. Affiliation to a totem must be represented visually, by a color, by a pattern or by a reduced image of the totem painted or tattooed on the skin or the belongings of the respective group. Presenting a totemic sign in public indicates communality for insiders and enforces distance to outsiders, deters enemies and invigorates allies, and challenges boundaries and affirms claims.

Totemism is by no means just a phenomenon of simple societies. Coats of arms and flags, but also the signs of idolized brands or sports clubs that are the object of contemporary merchandizing testify to the importance of totemic icons in complex societies. Profane objects such as t-shirts are saturated with sacrality when they bear the printed symbol or name of a cult brand, team or star.

The meaning of Totemism eludes utilitarian thinking. Far from being just an indication of membership in strategic cooperation, the presentation of totems amounts to a fundamental representation of collective identity; we belong to them, we are proud of it, and we show our power in the act of presenting the icon of our totem in public. Totems are not a profane means of attaining higher goals. Rather, they are sacred symbols. Because of their sacred nature, they cannot be easily exchanged or altered. The Christian cross cannot, for convenience's sake, be replaced by a red triangle, for example, and even instituting a change at the level of commercial brands' icons requires extensive public campaigning if it is to be successful at all. Totemic symbols are not just conventions. They partake instead in the sacred nature of collective identity and their presence sacralizes the site at which they are presented. The visible symbol and the invisible reality are not yet separate from one another.

Whereas totemic signs represent an invisible and sacred collective identity, images represent a reality that, while in principle visible, is nonetheless absent or gone. Images can represent humans, animals, or gods. Few simple societies, however, have already developed this pictorial representation of external objects that presupposes a certain awareness of the difference between representation and reality: the picture of a person is not the person him- or herself, but a representation of him or her. People in simple cultures do not yet distinguish strictly between the symbolic representation and the represented reference—when we mention the devil's name we invoke his presence. In this respect, they are no different from our everyday "common sense" attitude that takes the equivalence of sign and signified object for granted, only questioning this relationship on exceptional occasions. (Although the prehistoric cave paintings of animals and hunters in Europe and Africa seem to be a case in point, our knowledge about the cultural context of these paintings and drawings is quite limited.) The pictorial representations produced by simple societies consist mostly of sculptures representing ancestors or masks required for ritual dances. Their members possess little awareness of the difference between the symbolic representation and its referential reality. The statue of the ancestor is fused with the social meaning of the ancestor: it commands respect and cannot be misused for profane purposes. Having the statue of the ancestor with you and putting it at the center of your dwelling ensures the presence of the ancestor and enables you to ask for his support and protection, etc. But again: there is not much difference between the hunter carrying a little statue of an ancestor and the modern manager who puts photos of his family on his office desk or positions a painted portrait of his mother as the centerpiece of his living room.

Possibly the most important mode of representation in simple societies is the ritual performance. Ritual performances such as marriages or sacrifices are not just events, but iterations of events. They represent and repeat events that have occurred previously. Only by this reference to the past does the ritual become visible as a stereotyped and formalized practice. This standardization and formalism lies at the core of the ritual process. The ritual performance follows a regime, it refers to a script, it recalls previous events, and it transcends the reality at hand. However, the performance is never completely controlled and instructed by this script; it is never just the text and nothing more. It is also a concrete, visible, and particular event that catches the attention of those who attend it. In this merging of the particular event and the ritual pattern, in the iteration of events, the ritual constructs meaning. The meaning of an event results from its reference to another event, which, in turn, also draws its meaning from the reference to other events and so forth.

By presenting a surprising event as an iteration of previous events, the ritual performance enables participants to obscure the frightening impact of this event. Thus, it softens the edge of suddenness. It endows us with the capacity to cope with the extraordinary as something that has happened previously. Rituals perform an order; in relating events, they construct, for the moment of performance, an insurmountable and immutable reality.

Rituals differ from mere habits, although they share with them the strong relation to embodiment and corporeality. Habits cannot be wrong—they are simplified embodied regular behavioral sequences that do not conform to rules and are therefore not necessarily subject to judgments as to whether the performance was correct. Unlike habits, the performance of a ritual can be criticized for deviating from the rule and, therefore, for failing to create reality. As a rule-governed activity, ritual can never be confined to a single individual, whereas a habit can be displayed by an individual actor alone. Following a rule presupposes the perspective of an observer who is able to judge mistakes and correct the performance because he or she is a competent member of the respective community.

In their most elementary form, rituals do not just describe or imitate an order of the external world that is also available through other representations. Instead, the ritual performance creates the order and this order exists only because it is performed. When we dance—and dancing is a very important and almost universal ritual—we do not consider the performance of the dance to be nothing more than a weaker imitation of some external reality. Instead, the performance *is* the reality and the dance exists only insofar as it is performed. People who marry cannot, later on, invalidate the marriage by simply claiming that they were just playing bride and groom when they solemnly declared "I do" before their families and the representative of the community—it is by virtue of this very declaration that the act of marriage is performed.

It is the solemn and serious attitude of the participants that sets the ritual performance apart from the "non-serious" process of rehearsing, learning, or quoting a procedure or ceremony. The same bodily procedure can be framed as a play, as a rehearsal, as a mockery, as a quotation, or as a serious ritual performance. Thus, the ritual performance results neither from the inner attitude of the individual actor nor from the bodily behavior alone. Instead, it is generated by the collective definition of the performance as *serious*. The ritual is not just the expression of individual feelings or of a shared belief. Instead, it is the other way around: the performance of the ritual produces these feelings on the part of individual participants. Contemporary psychotherapeutic techniques like psychodrama use this power of rituals: they repeat a ritual setting in order to explore the emotional relations between the persons played in the performance.

Social reality can never entirely dispense with these ritual foundations. There must be an ultimate horizon where we simply give up our doubts as to whether the action is staged or spontaneous, artificial or authentic, or true or false and where we ignore any further questions and accept the performance as real.

Rituals, whether sacred or not, differ strongly from techniques or from the logic of "work." Techniques refer to an external reality that is available through other descriptions and that can be improved and criticized with reference to this external reality. In this respect, rituals also differ from games or plays that must follow certain rules. These rules can—under certain circumstances—be modified or partially replaced by other rules without risking a disruption of the performance of the game. If we are playing a game, we do not only know the difference between rules and rule-guided actions, but we perceive the other players from a strategic point of view. We assume a difference of interests, different roles, and different options to respond to the actions of others, etc. In playing a game, we are also able to fake an intention, to be suspicious, to criticize a strategic move, and to improve our performance. In contrast, rituals suspend and disregard the differences and divisions between the participating actors. They construct the elementary "communitas" (Turner 1989) that transcends social cleavages and unites the community before social order and social structure can take over. Beyond the bounds of constitutive ritual there exists only a frightening reality that is devoid of any form—the sacred or the demonic in its pure and unbearable existence.

> Field experience and general reading in the arts and humanities convinced me that the social" is not identical with the "social-structural." There are other modalities of social relationship.
>
> Beyond the structural lies not only the Hobbesian "war of all against all" but also communitas, a mode of relationship already recognized as such by our seminar. Essentially, communitas is a relationship between concrete, historical, idiosyncratic individuals. These individuals are not segmentalized into roles and statuses but confront one another rather in the manner of Martin Buber's "I and Thou." Along with this direct, immediate, and total confrontation of human identities, there tends to go a model of society as a homogeneous, unstructured communitas, whose boundaries are ideally coterminous with those of the human species. Communitas is in this respect strikingly different from Durkheimian "solidarity," the force of which depends upon an in-group/out-group contrast. To some extent, communitas is to solidarity as Henri Bergson's "open morality" is to his "closed morality." But the spontaneity and immediacy of communitas—as opposed to the jural-political character of structure—can seldom be maintained for very long. Communitas itself soon develops a structure, in which free

relationships between individuals become converted into norm-governed relationships between social personae.

Turner, Victor W. (1989): *The Ritual Process. Structure and Anti-Structure*. Ithaca: Cornell University Press, p. 131f.

This fundamental nature of rituals is rooted into their corporality. Rituals do not exist, as do texts or institutions, as structures of signification or dispositions of power and control. Instead, they exist as embodied performances, as events produced and experienced bodily by actors in a shared situation and at a local site. They consist entirely of bodily procedures—singing, dancing, moving, and marching, sometimes of painful violations of the body or even of killing a living being in a sacrificial rite.

In the ritual coordination of bodily movements, the participants experience themselves as mirrored by others—individual differences are disregarded in and even banned from the ritual performance. The ritual opens up a space for collective identity in its most elementary form. There is also almost no separation between the acting person and his or her role—both are merged in the ritual performance. Thus, the ritual performance produce a collectively shared, embodied reality that, is beyond any question or doubt, exempted from fluidification and negation, but also no longer driven by the imperatives of nature.

This feeling of authenticity is enhanced by the violent event that lies at the core of many ancient rituals. Ritual frames sublimate and purify the archaic violence that lurks behind the surface of everyday life. Instead of ignoring, denying, or repressing this violence entirely, the ritual extracts the violence from the members of the community, shifts it to an isolated and regulated space, blames it on a scapegoat and frames it within its opposite—order (Girard 1979). Just as the shamanistic practice removes the evil from the individual person and bans it to a particular object, the sacrificial ritual concentrates this dangerous violence on the victim and thus purifies the community. When the bloody sacrifice of the victim or the ritual fight between antagonists is finished, the audience can expect after this cathartic release to leave the site with a feeling of deep relief and even of internal peace.

All rituals are collective endeavors that draw a boundary between inside and outside. They achieve their construction of collective identity most effectively by excluding outsiders. Though similar in this exclusive and defensive attitude, rituals differ in their sensitivity to the actual presence of outsiders, that is, of persons who do not participate in the ritual process.

Solemn ceremonial performances that permit us to address the sacred are most sensitive to the presence of outsiders. Outsiders can destroy the potential of rituals to summon up the presence of the sacred. They do not know how to contribute to the process of ritual action and their very presence seems to contaminate the aura of the sacred. Outsiders who refuse to participate in the ritual process represent the regular and profane order of everyday life—they prevent the generation of the ritual communitas (Turner 1989). For this reason, solemn ceremonial rituals try to exclude these profane outsiders physically or socially from the locality. The performance of the ritual is shielded from the eyes and ears of outsiders or their presence is bound to strict rules of respect that enable the ritual actors to ignore them.

In distinction to these profane outsiders, there is also the sacred outsider who participates in and even structures the ritual process. Priests, monks, and officials are cases of such figures who are stripped of normal social, sexual, or political attributes and who, by their very exemption from a position in the regular and profane order, are able to construct liminality (Turner 1989). These sacred outsiders are not involved in conflicts and feuds. They are not moved by retaliation or revenge. They represent a thirdness that breaks down the profane order, opening a space for communitas and for the encounter with the sacred.

A special case of this thirdness is the Homo Sacer (Agamben 1998) who is ceremonially sacrificed in order to constitute the social order. This sacrificial constitution of the social order presupposes that the Homo Sacer is an outsider, an individual who is exempted from the profane social order. If the Homo Sacer had a position within the kinship structure, his death would inevitably engender a vicious circle of violent revenge and retaliation. The family of the killed person would respond violently even if this person were considered guilty of a crime. Because the attribution of guilt requires an assumption of norms and obligations that are valid only for members of the community, the sacrificial victim must be innocent. Thus, the ceremonial killing of the innocent outsider prevents or stops the circle of violence within the community (Girard 1979).

In a similar manner to ceremonial rituals' sensitivity to the presence of profane outsiders, they also must overcome disorder and evil by banning, expelling, defeating, or destroying these powers. Most ritual communities are haunted by the fear that evil outside their boundaries could transgress them, invade the bodies of community members and be present within the community, albeit in a disguised way. This is the origin of beliefs in witchcraft, in devils, soul stealers, and body snatchers. Spotting them answers the question why evil is affecting a person or a community—in contrast to the question how evil is operating (which is at the center of the scientific perspective on evil).

When the source of evil is identified rituals of purification, exorcism, and sacrifice try to remove, ban, or expel the impure, the demonic, and the evil from the body of the ritual community. Because the ritual *constitutes* reality, the opposite of order may only be performed and embodied to the end of being overcome and destroyed in an exemplary fashion. If it could not be expelled or purified, this evil would threaten and destroy the social order. It is for this reason that a ritual performance—in contrast to theatrical performances—cannot have a tragic ending.

References

Agamben, Giorgio (1998): *Homo Sacer. Sovereign Power and Bare Life*. Stanford: Stanford University Press.

Bloch, Marc (1973): *The Royal Touch. Sacred Monarchy and Scrofula in England and France*. London: Routledge & Kegan Paul; Montreal: McGill-Queen's University Press.

Douglas, Mary (1966): *Purity and Danger. An Analysis of Concepts of Pollution and Taboo*. New York; Washington: Frederick A. Praeger.

Durkheim, Émile (1965): *The Elementary Forms of Religious Life*. New York: Free Press.

Eliade, Mircea (1959): *The Sacred and the Profane. The Nature of Religion*. New York et al.: Harcourt.

Enzensberger, Hans Magnus (1994): *Civil Wars. From L.A. to Bosnia*. New York: The New Press.

Evans-Pritchard, Edward Evan (1976): *Witchcraft, Oracles, and Magic Among the Azande*. Oxford: Oxford University Press.

Girard, René (1979): *Violence and the Sacred*. Baltimore: Johns Hopkins University Press.

Lee, Richard B. and Irven DeVore (1972, eds.): *Man the Hunter*. Chicago: Aldine Atherton.

Lévi-Strauss, Claude (1969): *The Elementary Structures of Kinship*. Boston: Beacon Press.

Lévi-Strauss, Claude (1976): *Tristes Tropiques*. Harmondsworth: Penguin Books.

Malinowski, Bronislaw (1984): *Argonauts of the Western Pacific. An Account of Native Enterprise and Adventure in the Archipelagos of Melanesian New Guinea*. Prospect Heights: Waveland Press.

Malinowski, Bronislaw (1989): "The Problem of Meaning in Primitive Languages". In: Ogden, C. K. and I. A. Richards (eds.): *The Meaning of Meaning. A Study of the Influence of Language upon Thought and of the Science of Symbolism*. San Diego et al.: Harcourt Brace Jovanovich Publishers, 296–336.

Mauss, Marcel (1990): *The Gift. The Form and Reason for Exchange in Archaic Societies*. New York: W.W. Norton.

Radcliffe-Brown, A. R. (1979): *Structure and Function in Primitive Society. Essays and Addresses*. London; Henley: Routledge & Kegan Paul.

Sahlins, Marshall D. (1972): *Stone Age Economics*. Chicago: Aldine-Atherton.

Service, Elman R. (1979): *The Hunters*. Englewood Cliffs: Prentice-Hall.

Turner, Victor (1989): *The Ritual Process. Structure and Anti-Structure*. Ithaca: Cornell University Press.

Van Gennep, Arnold (1960): *The Rites of Passage. A Classic Study of Cultural Celebrations*. Chicago: University of Chicago Press.

Weber, Max (1978): *Economy and Society. An Outline of Interpretive Sociology*. Berkeley: University of California Press.

The Transition to a Sedentary Life

Bernhard Giesen

Production: Horticulture and Herding

The first breakthroughs that enabled the sustenance of larger communities consisted in the planting of crops and in the breeding of cattle. Some 8000 years ago, the first societies changed their major mode of production from collecting what was available in the environment to producing food by themselves in their immediate surroundings. This was, of course, not a matter of deliberate decision and planning, but a gradual transition that occurred over a long period of time. Today, the societies of the Yanomanö, Bororo, Jivaro, or Mundurucu in tropical America or the ethnic groups in the New Guinea Highlands are still based on this ancient mode of production. Horticulture begins when a group occasionally plants wild fruits such as bananas, manioc, potatoes, or grain in little gardens around their village and the plants grow to yield crops. This shift from collecting to controlled production engenders far-reaching changes in the way of life in simple societies. Planting and harvesting crops presupposes a continuous residence at the place where they grow—not only so as to enjoy the results of the planting efforts but also to protect the plants against animal or human foraging.

Turning from a nomadic to a sedentary lifestyle allows groups to build large and durable houses, to store goods and food, and to acquire a certain modest wealth that could potentially be stolen by people from other villages and hence must be protected. Sometimes, low fences are constructed

B. Giesen et al., *Human Societies*, Cultural Sociology,
https://doi.org/10.1007/978-3-032-07697-7_4

to prevent wild animals from ravaging the garden. The simple and light tools of the hunters and gatherers are replaced by additional ones such as hoes and digging sticks. Still, these tools are not yet manufactured from metal and irrigation is unknown.

In tropical areas, the soil is frequently exhausted after a few harvest cycles. The village must therefore move to a new location, fell the trees in the forest, use the wood as materials for new houses, and burn the remaining wood, the ashes of which serve as fertilizer for the new fields. This slash-and-burn horticulture allows only for a limited degree of sedentary existence in the tropical forests and increases the chances of conflicts between an old established village and a newly settling one. Horticulture yields food for a larger population on a given territory than does hunting and gathering. The average size of the village is much larger than in simple societies, but even in villages amounting to a thousand persons or more, the societal community consists of people who could meet each other face to face. As in simple societies, production of food exceeds the subsistence level only on exceptional occasions. Horticulturalists still live in the original affluent society and generally enjoy extensive leisure time, particularly as far as the males are concerned. Gardening is usually women's work; this shift of responsibility for the primary production of food to women has consequences that we will deal with in the next chapter.

Horticulture frequently goes hand in hand with the domestication of animals such as dogs, fowl, or pigs that live among the huts and houses in the village. Here again, food is no longer collected by roaming through the plains or forests, but produced by the villagers at home. The main source of food in herding societies is, however, meat they produce themselves through the breeding of domesticated cattle, sheep, camels, or horses. Many herding societies ban hunting as a mode of food production. As strict as their distinction toward hunters may be they do not, however, shift completely to a sedentary way of life. Like the slash-and-burn horticulturalists, they must respond to the exhaustion of the fertile soil on which their food source depends. Thus they wander with their animals from one grazing ground to the next and organize their life around the distance to the next waterhole or river.

Many herding societies separate the male members who move around with the herd from the women who stay in the village, care for the children, and prepare for the return of the herdsmen. Breeding cattle, horses, or sheep allows both for mobile storage and for increasing wealth beyond the level of subsistence. Owning a large number of cattle is, however, not regarded as an opportunity for trading in other goods, but seen as a matter of honor and pride. A man's social standing and reputation stem not only from his valor as a warrior, but also from his wealth of cattle and sometimes both are interlinked. Cattle is also an important bride wealth and stealing it can become a major issue of vengeance and retaliation. Today, the societies of the Dinka, Somali, Maasai, and other ethnic groups in East Africa, the Tuareg in Central Sahara, and Central Asian ethnic groups still live by this herding mode of production.

Classification: Gender Separation, Male Association, and the Fear of Witchcraft

The shift of the main role of production from male hunting to female gardening has a strong impact on social classification. With respect to kinship, it amounts to a change in the construction of lineage and descent. In many horticultural societies, patrilinearity and patrilocality, the dominant mode of constructing lineage in simple societies, are replaced by matrilinearity and matrilocality. It is the woman's family that names children and provides the locality where the family lives. Here, the mother is the stable center of the family, whereas the father can be a relatively rare visitor who mostly roams about with his companions. The strong attachment to the mother as the core of the family is reflected by the sons' outraged emotional response to insults directed at her. Degrading jokes or insults aimed at the father do not provoke similar responses. Matrilocality decreases the importance of the father for the family. The uncle frequently is attributed relatively greater importance in such societies. He becomes the male figure that protects the family and takes revenge if one of its members is harmed.

Horticultural and herding societies separate adult men and women much more stringently than any other type of society. While women represent the center of the family, adult men often gather together in special places like the long houses of New Guinea to which only males are admitted. Here they drink and smoke, boast about their courageous exploits and feats, or remember killed relatives who call for revenge. Thus they vie with one another in a special male type of communication, challenging each other's courage and valor. Challenges and statements of defiance are launched and accepted, claims to courage and valor are put to test, and, in the end, the male group is prepared to kill a dangerous animal or to assault a neighboring village, kill some of its men, and abduct its women. The war party leaves the familiar environment, ventures out to unknown territory and, beyond the confines of regular life, engages in extraordinary and dangerous actions.

> Although few raids are initiated with the sole intention of capturing women, that is always a desired side benefit. But a few wars are started with the intention of abducting women. One village in Brazil that I visited in 1967 had a critical shortage of women. Some missionaries who lived in the village learned later of the treacherous means by which the group alleviated its problem and gave me this account. A raiding party organized by the group's headman went to a distant village and told the people there that they had received machetes and cooking pots from foreigners who prayed to a spirit for such items. The raiders volunteered to teach these people how to pray to the same spirit. When the men knelt down and bowed their heads, the raiders attacked and killed them with machetes, took their women, and fled. Treachery of this kind, called *nomohori*—dastardly trick—is the ultimate form of Yanomamö violence. Kaobawä's group had suffered a treacherous massacre in 1950, as described in the introduction and in chapter 5, although in that case the treachery was in revenge for an earlier killing. Still, the assailants did attempt to abduct women after the objectives of their treachery were accomplished. In fact, had it not been for their greed to capture women, the massacre would have been even more successful.
>
> Chagnon, Napoleon A. (1992): *Yanomamö. The Last Days of Eden*. San Diego: Harcourt and Brace, p. 218f.

In these raids, animals and humans are killed in an archaic thirst for bloodshed. The difference between dangerous animals and unknown strangers is perceived as minor. For the raiders, strangers are not members of the community of humans, but they are also not mere profane objects that can be used and eaten like regular hunting prey. The rise of archaic warfare is not driven by economic motives, nor is it triggered by a quest for revenge for harm inflicted on one's own family. Instead, it is motivated by a new attempt to mark the distinction between the sacred and the profane. Killing a lion or a male warrior of another tribe confirms and corroborates the power of our own sacred core, our own sovereign power over the life and death of a beast that is demonically sacred in itself. Here, a new distinction is necessary for what was, in less complex cultures, viewed as an unclear liminal existence between the sacred and the profane. This realm is now classified as the demonic opposite of the sacred. It is neither sacred in the positive redeeming sense of the term nor is it profane. Thus, archaic warfare represents the attempt to mark the boundary between the sacred and the demonic that was not yet salient or still blurred in simple societies. Since it is not just a profane object, the slain demonic beast cannot be consumed. Instead, its remains are turned into trophies and proudly displayed at home in order to prove the victory over the beast. This victory of the sacred as represented by the hero over the monster, the beast, the enemy is the cultural motive of archaic warfare.

Retaliation for previous raids and the drive to demonstrate heroism are presented as the main motives of the raiders. Thus a vicious circle of raid and counter raid can be established that results in high death tolls. Some horticultural societies were among the most violent and aggressive societies in human evolution:

> (1) Approximately 40 percent of the adult males had participated in the killing of another Yanomamö. The majority of them (6o percent) had killed only one person, but some were repeatedly successful as warriors and had participated in the killing of up to sixteen.
>
> (2) Approximately 30 percent of all the deaths among adult males were due to violence.
>
> (3) Approximately two-thirds of the people aged forty or older had lost through violence at least one of these close biological relatives: a parent, a sibling, or a child. Most of them (57 percent) had lost two or more such close relatives. That helps to explain why large numbers of the people are motivated by revenge, and why kinsmen keep gourds containing the ashes of slain heroes for many years.
>
> The most unusual and impressive finding, one that was subsequently discussed and heatedly debated in the press and in academic journals, was the correlation between military success and reproductive success. *Unokais*—men who have killed—were more successful at obtaining wives and, as a consequence, had more offspring than men the same age who were not *unokais*. The *unokais* had, on average, more than two and a half times as many wives as non-*unokais* did, and more than three times as many children.
>
> Chagnon, Napoleon A. (1992): *Yanomamö. The Last Days of Eden.* San Diego: Harcourt and Brace, p. 239f.

The rise of archaic warfare is not only due to a special mode of male communication and driven by the attempt to mark the difference between the sacred and the demonic. It is also fostered by a new type of association that runs counter to the lines of kinship. Membership in these warrior associations is reserved for adult males and based on cooptation. In contrast to family ties that appear as naturally given, these male associations must construct ties of solidarity through special rituals and cultural codes of honor. Whoever has the honor of being admitted into these associations or cultural brotherhoods as a member must undergo a special ritual and is obligated to uphold solidarity, obedience, and peace within the association, in contrast to the violence and distrust that reign in relations with outsiders. Membership in these associations is often secret, invisible for outsiders but noticeable for insiders. These secret societies are a new mode of organization that transcend the natural ties of kinship and the volatile ties of charismatic leadership. They allow for the inclusion of a large number of adult males without relying on conquest and hierarchy.

At the same time, they produce a new type of conflict. The unconditional commitment toward the secret brotherhood can collide with the obligation of family ties, a conflict which is extremely difficult to resolve. Disobeying your father, leaving your family, and following a charismatic leader or executing the commands of the brotherhood is a demanding solution that, in principle, breaks the structural backbone of simple societies. For this reason, such actions lie at the core of many mythical conflicts. Though male associations originated in horticultural societies, they are by no means confined to this mode of social organization. We know of many variations of this pattern ranging from the Leopard societies in central Africa to the followers of Jesus Christ ("Give up your brothers and sisters and follow me"), from medieval monastic movements to mafia organizations, from old boys' associations to terrorist networks. As different as they are, they converge in their departure from family ties and local roots. The members of these associations are committed to a cause that transcends not only kinship, but also locality. Local situations and translocal networks, everyday normality and extraordinary commitment, and the obvious world at hand and the secret society out there are separated. The inside-outside distinction is completed by the distinction between translocal secrecy and obvious local embeddedness.

The concept of a powerful secret association that pulls strings and punishes the disobedient also has its demonic counterpart in the suspicion of witchcraft and sorcery that lurks behind the surface of everyday common sense. While simple societies do not strictly distinguish between the sacred and the demonic, as we have outlined above, the fear of demons is of paramount importance for people in horticultural societies. For them, the world is not only moved by the distinction between the sacred and the profane, but the sacred is also dually layered. There is the mundane and visible reality where things appear as what they are and then there are invisible powers beyond the surface of things and these invisible powers are split into the good and the evil ones, the helpful and sacred spirits, and the destructive and demonic ones. These demonic powers can creep into familiar appearances, bewitch and possess human beings, steal their soul, and turn them into dangerous creatures. Evil is no longer located simply beyond the boundaries of the community, in the wastelands outside, but instead it can secretly invade the community. Whereas the members of the community are aware of the location of the positive sacred and recognize its presence, the demonic, by contrast, is invisible to those who are threatened by it. Danger is defined by uncertainty and risk. At a superficial glance, everything might seem to be as it should, but this semblance cannot be trusted. It is wise to be permanently suspicious of hidden evil, of witchcraft, of poison, and of secret conspiracies. Like dirt or waste, the invasion of the demons amounts to a fundamental blurring of cultural boundaries. But unlike dirt that is visible and must be concealed or removed, the demons are invisible and cannot be discovered without special procedures. This invisibility increases the danger. If nobody is aware of the demons they can do their evil work without restraint. For this reason, sorcery and witchcraft require counteractive magic, for example protection by amulets and experimental rituals that reveal the true nature of things. We will return to these ritual procedures at the end of this chapter. The suspicion of hidden danger as a cultural motif is, of course, not confined to horticultural societies. We also encounter this attitude in traditional and modern societies. Witch-hunts and anti-Semitism, fear of viruses, poisoning, or invasions of extraterrestrial aliens—as different as they are—are moved by the same motive to discover the evil that has secretly invaded the community.

Horticultural and herding societies classify their population not only by kinship and membership in translocal associations. In contrast to simple band societies, they distinguish institutionally between age groups such as children, young marriageable men and women, adult persons, and old people. These age groups are clearly separated by appearance (hair style, dress, and decoration), social status (rights, access to ritual places, and family position), and communication (inside vs. outside). The transition of a person from one age group, for example, childhood, to another, for example, young warriorhood, does not occur gradually, but instead is accomplished by means of an elaborate ritual of passage which is occasionally accompanied by a change of name. This strict separation between age groups creates in turn clearly separated memories: every age group has a collectively shared memory that differs from other age groups, creating a difference that requires mediation. A first vague sense of historical time differences arises.

Circulation: Rituals of Initiation, Sacrifices, and Feasts

Circulation in horticultural and herding societies is mostly based on the elaboration of patterns that can already be found in simple band societies. The principle of diffuse reciprocity, for example, is sometimes used to construct large rings in which a ceremonial gift is passed sequentially from one group to another. The circular motion of a ceremonial gift ensures peace and demonstrates the mutual respect enjoyed by those participating in this institutional pattern. The most famous of these institutions is the Kula ring of seafaring societies in the Pacific (Malinowski 1984). In contrast to the gift that is passed from the leader to his followers, the ceremonial ring establishes peaceful relations among members who consider themselves equal in political power and social rank.

> The Kula is a form of exchange, of extensive, inter-tribal character; it is carried on by communities inhabiting a wide ring of islands, which form a closed circuit. [...] Along this route, articles of two kinds, and these two kinds only, are constantly travelling in opposite directions. In the direction of the hands of a clock, moves constantly one of these kinds—long necklaces of red shell, called *soulava* [...]. In the opposite direction moves the other kind—bracelets of white shell called *mwali* [...]. Each of these articles, as it travels in its own direction on the closed circuit, meets on its way articles of the other class, and is constantly being exchanged for them. Every movement of the Kula articles, every detail of the transactions is fixed and regulated by a set of traditional rules and conventions, and some acts of the Kula are accompanied by an elaborate magical ritual and public ceremonies. On every island and in every village, a more or less limited number of men take part in the Kula—that is to say, receive the goods, hold them for a short time, and then pass them on. Therefore every man who is in the Kula, periodically though not regularly, receives one or several *mwali* (arm-shells), or a *soulava* (necklace of red shell discs), and then has to hand it on to one of his partners, from whom he receives the opposite commodity in exchange. Thus no man ever keeps any of the articles for any length of time in his

possession. One transaction does not finish the Kula relationship, the rule being "once in the Kula, always in the Kula," and a partnership between two men is a permanent and lifelong affair. Again, any given *mwali* or *soulava* may always be found travelling and changing hands, and there is no question of its ever settling down, so that the principle "once in the Kula, always in the Kula" applies also to the valuables themselves. [...] The Kula is thus an extremely big and complex institution, both in its geographical extent, and in the manifoldness of its component pursuits. It welds together a considerable number of tribes, and it embraces a vast complex of activities, interconnected, and playing into one another, so as to form one organic whole.

Malinowski, Bronislaw (1984): *Argonauts of the Western Pacific. An Account of Native Enterprise and Adventure in the Archipelagoes of Melanesian New Guinea*. Prospect Heights: Waveland Press, pp. 81–83

A special mode of gift-giving that can be found only in very crude forms in simple societies is the sacrifice. Here, too, the meaning of the gift is to establish a diffuse reciprocity. However, in this case, the receiver of the gift is not another human being, but a supernatural power, the ancestors, or, in most cases, a god or a goddess. In order to please the gods, a precious object is offered to them, a young animal is slaughtered, food is presented, decorative items are donated, or promises are made. Sacrificial items may not be polluted by profane use and sacrificing them must be expensive for the giver. Instead of offering an old and useless object that the giver wants to get rid of, a precious and cherished item such as a young animal or—in the case of the archaic human sacrifice—an innocent young person must be offered to the deities. We will recall that Abraham was expected not to sacrifice an old sheep, but his own son.

A few years ago a colleague of mine was travelling in Africa by boat and ran into a storm that seemed to grow dangerous. Suddenly a fellow passenger, a politician of some rank in that country, began throwing dollar bills into the raging waters. While we may share my colleague's astonishment, the very metaphor of "raging waters" shows how easily we let ourselves slip into personifying the natural forces.

Similar astonishment and plain ridicule of such behavior can be found in some ancient sources, at least among the philosophers. Seneca, in his *Naturales Quaestiones*, writes:

I do not hold myself back from revealing all the absurdities of our people. They say certain persons are experienced in observing clouds and capable of foretelling when there will be hail. ...It's really impossible. At Kleonai there were official "hail-watchers" (*chalazophylakes*), watching for hail to come. If

> they said that hail was coming, what do you think the people did? Each person would sacrifice individually: one a lamb, the other a chicken; at once those clouds moved away to some other place, when they had savored some blood. You laugh? Listen to what will make you laugh even more. If a man had neither lamb nor chicken, he sacrificed what he could afford: he laid hands on himself—but don't think clouds are gluttonous or cruel. He punched his finger with a very sharply pointed pen, and with this blood he performed his auspicious sacrifice; and the hail turned away from his piece of land as well as from places where they had been implored with greater sacrifices.
>
> Seneca adds that the hail-guardians were sued and punished if they failed to avert disaster from vineyards and cornfields. The rationalist laughs at this response to panic because it shows no obvious link between means and ends, especially in the face of natural forces. The reaction of panic is to give up valuables, to kill one's own animals, to inflict wounds on oneself. Seneca does not hesitate to speak of sacred actions, sacrifice (sacrificare). At Kleonai this was a well-established, institutionalized religious ritual.
>
> Burkert, Walter (1998): *Creation of the Sacred. Tracks of Biology in Early Religions.* Cambridge: Harvard University Press, p. 34f.

The sacrifice cannot directly force supernatural powers to bless the giver with grace and support, but it will please the gods and increase the chance of grace and mercy. In many cases, after being presented and offered to the deities, the sacrificial animal or crops will be consumed by the priests or the people in a ceremonial meal. Here, the common meal will strengthen the ties between the participants. Commensal rituals such as a dining together on festive occasions symbolize collective identity and contribute to social solidarity. It is difficult to kill somebody with whom one has just shared a meal.

The passage from one age class to another follows the general logic of rituals of passage outlined above, but it is object of much more complex ceremonies. Those who are to be initiated into the class of young warriors must undergo complicated procedures of purification and tests of their bravery. They are assembled in special camps outside of the village, must endure bodily tortures, are subjected to starvation and isolation, and placed in a liminal situation in which dream and reality are fused. After passing these tests, they are ceremonially incorporated into the new status group, invested with the signs of their new dignity, and celebrated by the community. The logic of these rituals of initiation is very similar to that underlying the procedures of academic examination and graduation.

Horticultural and herding societies are based on the ability to accumulate and store a certain amount of food. This creates the potential threat that outsiders might steal cattle and crops. While crops are too cumbersome to transport over longer distances, animals can be driven away more quickly. Therefore, the protection of the herds is a core concern in these societies. Although stealing the livestock of another tribe is not considered to be unethical, it is not part of the regular mode of circulating goods and it is not the prime motive for warfare.

An important incentive for surplus production in horticultural societies are the feasts to which people of neighboring villages are invited. Occasionally, men wish to boast of their prowess and organize large ceremonial celebrations to impress the invited guests and even to invoke their jealousy. In order to fund these celebrations and to feed the guests, the members of the village must cross the limits of subsistence production—pigs and cattle are slaughtered, hunts must yield additional game, stock and stores of crops must be emptied. Thus, for a limited time, the leisurely life of the original affluent society is transformed into a collective effort to produce more than required for mere subsistence. In many cases, this collective drive for surplus production is stimulated by a charismatic leader, an important man who urges his followers to impress the guests and to gain their admiration. The guests, in turn, feel pressure either to recognize the superiority of the hosting group and to admit to it or to counter the challenge by holding their own, at least equally impressive celebration. In this way, generosity and charismatic leadership merge with male competition and group distinction. As a result of this process, the organization of festive celebrations can become a regular institution that requires surplus production on a relatively continuous level. The competitive nature of the endeavor overcomes the limiting barrier of subsistence production. More work yields more food, more food enables the group to organize more celebrations, and more celebrations engender more recognition. But these celebrations do not only stimulate surplus production but they also serve as a payment of debts that have been accumulated for some time.

> The economic system of the Indians of British Columbia is largely based on credit, just as much as that of civilised communities. In all his undertakings the Indian relies on the help of his friends. He promises to pay them for this help at a later date. If the help furnished consisted in valuables, which are measured by the Indians by blankets as we measure them by money, he promises to repay the amount so loaned with interest. The Indian has no

System of writing, and therefore, in order to give security to the transaction, it is performed publicly. The contracting of debts, on the one hand, and the paying of debts, on the other, is the potlatch. This economic system has developed to such an extent that the capital possessed by all the individuals of the tribe combined exceeds many times the actual amount of cash that exists; that is to say, the conditions are quite analogous to those prevailing in our community: if we want to call in all our outstanding debts, it is found that there is not by any means money enough in existence to pay them, and the result of an attempt of all the creditors to call in their loans results in disastrous panic, from which it takes the community a long time to recover.

It must be clearly understood that an Indian who invites all his friends and neighbours to a great potlatch, and apparently squanders all the accumulated results of long years of labour, has two things in his mind which we cannot but acknowledge as wise and worthy of praise. His first object is to pay his debts. This is done publicly and with much ceremony, as a matter of record. His second object is to invest the fruits of his labour so that the greatest benefit will accrue from them for himself as well as for his children. The recipients of gifts at this festival receive these as loans, which they utilise in their present undertakings, but after the lapse of several years they must repay them with interest to the giver or to his heir. Thus the potlatch comes to be considered by the Indians as a means of insuring the well-being of their children if they should be left orphans while still young. It is, we might say, their life insurance.

The sudden abolition of this system—which in all its intricacies is very difficult to understand, but the main points of which were set forth in the preceding remarks—destroys therefore all the accumulated capital of the Indians. It undoes the carefully planned life-work of the present generation, exposes them to need in their old age, and leaves the orphans unprovided for. What wonder that it should be resisted with vigour by the best class of Indians, and that only the lazy should support it, because it relieves them of the duty of paying their debts?

Boas, Franz (1974): *The Shaping of American Anthropology, 1883-1911: A Franz Boas Reader*, edited by George W. Stocking, Jr. New York: Basic Books, p. 105f.

Representation: Trophy Taking, Protection Against Evil Spirits, and Revelatory Rituals

The sedentary lifestyle of horticulturalists allows for larger material forms of representation. Masks and statues representing the ancestors or protective spirits can be oversized, wood carving and paintings can be used to decorate the houses. Relics of ancestors are enclosed in specially decorated containers and stored at special places that are exempted from profane use.

As important as the veneration of the ancestors is, it is not a mode of representation that sets horticultural societies apart from simple societies. More important in this respect is the cult of trophy taking that comes along with archaic warfare. Trophies are relics that represent the defeated enemy, the conquest of the wilderness, the successful encounter with danger or the killing of the demonic beast. They stand for the heroic effort to overcome the demonic outside that, after its defeat, could be moved inside and placed at the center of the home. Although the trophy has lost its intimidating demonic power it is not just a profane object. As a relic it tells a story of personal heroism and domination. The trophy takers defied risk and danger, ventured out to the unknown, met the challenge of wilderness, defeated the monster and returned home with a proof of what, later on, will be called their heroism. The widespread myth of the monstrous animal, of the dragon, the huge ape, the evil monster, etc. that demands human sacrifice and is defeated by the courageous young hero reflects this calling of heroism. The challenge of the monster represents the unclassifiable demon who is half sacred and half profane, half god and half animal. The hero, by contrast, is seen as the guardian of the community and as the warrant of its safety. Heroism conveys a strong sense of individual personhood and has provided an icon of subjectivity that can be found in many cultures since it was invented by horticultural societies. The hero is, by definition, the extraordinary and unique individual who is exempted from common rules and reason. He is seen as half human and half divine and thus embodies the sacred in a special way. And we have to admit that, even today, this myth has an attractive appeal to many young men.

> When the community passes from peaceable savagery to a predatory phase of life, the conditions of emulation change. The opportunity and the incentive to emulation increase greatly in scope and urgency. The activity of the men more and more takes on the character of exploit; and an invidious comparison of one hunter or warrior with another grows continually easier and more habitual. Tangible evidences of prowess—trophies—find a place in men's habits of thought as an essential feature of the paraphernalia of life. Booty, trophies of the chase or of the raid, come to be prized as evidence of preëminent force. Aggression becomes the accredited form of action, and booty serves as *prima facie* evidence of successful aggression. As accepted at this cultural stage, the accredited, worthy form of self-assertion is contest; and useful articles or services obtained by seizure or compulsion, serve as a

> conventional evidence of successful contest. Therefore, by contrast, the obtaining of goods by other methods than seizure comes to be accounted unworthy of man in his best estate. The performance of productive work, or employment in personal service, falls under the same odium for the same reason. An invidious distinction in this way arises between exploit and acquisition by seizure on the one hand and industrial employment on the other hand. Labour acquires a character of irksomeness by virtue of the indignity imputed to it.
>
> Veblen, Thorstein (1957): *The Theory of the Leisure Class. An Economic Study of Institutions.* London: George Allen & Unwin, p. 16f.

Trophy taking is, of course, not only a matter of archaic warfare. It is not limited to skulls, skins, and scalps or, later on, to displaying the flags and banners of the defeated enemy in memorial halls. The logic of trophy taking governs also the winning of cups and medals in modern sports or the purchase of souvenirs by tourists. The little stones and shells that we collected during a beach holiday, the rug that we bargained in Marrakech, even the plaster copy of Michelangelo's David that we bought in Florence embody the extraordinary event of venturing out to a foreign region. Disconnected from the extraordinary event these items would be banal and profane objects: if the copy of David is purchased by mail order instead of being bought in Florence by us it is not a souvenir, it does not tell a story, it could be dumped into the garbage can.

Trophy taking is an important and pervasive mode of representing the challenge of the demonic outside. More complex modes of representation apply if the demonic powers can secretly and invisibly invade the community. Against this danger we have to protect us by wearing objects that exude defensive power. Sometimes these amulets are relics, parts of totemic animals, or magical objects prepared by a shaman. Sometimes however the protective objects are just emblematic representations of totems or images of the sacred. Displaying the totemic emblem is to invigorate its carrier and to deter the evil powers. Again, this shielding by relics and emblems is not just a matter of horticultural societies. Even in contemporary societies images of saints can be found in vehicles, blessings are to protect against diseases, amulets are to ensure luck, sacred water is to purify those who bath in it, the sign of the cross or of yin and yang is to foster luck, etc. These relics, emblems, and rituals are not just habitually displayed or performed for decorative purposes. Instead, even today people really believe that they impact on their luck or protect against misfortune.

Protective devices are completed by ritual procedures that are to reveal and to discover the persons that are bewitched or demonized. Since witchcraft works secretly and invisibly these procedures are thought to be essential to spot the evil within the society. Sometimes discovering certain marks of evil is regarded as being sufficient, sometimes specialists are called on in order to submit the person in question to revelatory rituals such as tortures, oracles, or investigations. These revelatory rituals aim at provoking the demons who have invaded the body to speak out. It is not simply the arbitrary judgment of the ritual specialist but the proof presented by the bewitched person herself that provides the basis for the verdict. This proof producing revelatory rituals is, of course, not only a matter of shamans in horticultural societies: ritual exorcism was practiced even in modern societies and also special techniques of interrogation in modern police investigations continue the logic of revelatory rituals: the accused person himself or herself has to disclose the evidence and commit to the crime. The truth revealed by these rituals can be questioned only with respect to the proper performance of the ritual. In contrast, the power of the ritual to reveal the truth about the evil is unchallenged.

> The method of revealing what is hidden by administering poison to fowls has a wide extension in Africa. [...] The poison used is a red powder manufactured from a forest creeper and mixed with water to a paste. The liquid is squeezed of the paste into the beaks of small domestic fowls which are compelled to swallow it. Generally violent spasms follow. The doses sometimes prove fatal, but as often the fowls recover. Sometimes they are even unaffected by the poison. From the behaviour of fowls under this ordeal, especially by their death or survival, Azande receive answers to the questions they place before the oracle. [...]
>
> The poison oracle, *benge*, is by far the most important of the Zande oracles. Zande rely completely on its decisions, which have the force of law when obtained on the orders of a prince. A visitor to Zandeland hears as much of the poison oracle as he hears of witchcraft, for whenever a question arises about the facts of a case or about a man's well-being they at once seek to know the opinion of the poison oracle on the matter. In many situations where we seek to base a verdict upon evidence or try to regulate our conduct by weighing of probabilities the Zande consults, without hesitation, the poison oracle and follows its directions with implicit trust.
>
> No important venture is undertaken without authorization of the poison oracle. In important collective undertakings, in all crises of life, in all serious legal disputes, in all matters strongly affecting individual welfare, in short,

on all occasions regarded by Azande as dangerous or socially important, the activity is preceded by consultation of the poison oracle. [...]

Basically, the system of question and answer in oracle consultations is simple. There are two tests, the *bambata sima*, or first test, and the *gingo*, or second test. If a fowl dies in the first test then another fowl must survive the second test, and if a fowl survives the first test another fowl must die in the second test for the judgement to be accepted as valid. Generally the question is so framed that the oracle will have to kill a fowl in the first test and spare another fowl in the corroborative test to give an affirmative reply, and to spare a fowl in the first test and kill another fowl in the corroborative test to give a negative reply; but this is not invariably the case, and questions are sometimes framed in an opposite manner. [...]

In the two tests one fowl must die and the other must live if the verdict is to be accepted as valid. If both live or both die the verdict is invalid and the oracle must be consulted on the matter a second time on another occasion.

Evans-Pritchard, Edward Evan (1976): *Witchcraft, Oracles, and Magic Among the Azande*. Oxford: Oxford University Press, pp. 120–122, 138f.

References

Boas, Franz (1974): *The Shaping of American Anthropology, 1883-1911: A Franz Boas Reader*, edited by George W. Stocking, Jr. New York: Basic Books.

Burkert, Walter (1998): *Creation of the Sacred. Tracks of Biology in Early Religions*. Cambridge: Harvard University Press.

Chagnon, Napoleon A. (1992): *Yanomamö. The Last Days of Eden*. San Diego: Harcourt and Brace.

Douglas, Mary (1966): *Purity and Danger. An Analysis of Concepts of Pollution and Taboo*. New York; Washington: Frederick A. Praeger.

Evans-Pritchard, Edward Evan (1976): *Witchcraft, Oracles, and Magic Among the Azande*. Oxford: Oxford University Press.

Harner, Michael (1972): *The Jivaro. People of the Sacred Waterfalls*. Berkeley and Los Angeles: University of California Press.

Malinowski, Bronislaw (1984): *Argonauts of the Western Pacific. An Account of Native Enterprise and Adventure in the Archipelagos of Melanesian New Guinea*. Prospect Heights: Waveland Press.

Service, Elman R. (1975): *Origins of the State and Civilization. The Process of Cultural Evolution*. New York: W.W. Norton & Company.

Veblen, Thorstein (1957): *The Theory of the Leisure Class. An Economic Study of Institutions*. London: George Allen & Unwin.

Conquest Societies

Bernhard Giesen

Production: Irrigation, Agriculture, Metallurgy, and the Use of Animal Energy

The so-called Neolithic revolution that occurred in Egypt, Mesopotamia, and India about four to six thousand years ago changed the structure, institutions, and cultures of human societies more than any other evolutionary breakthrough later on. It was based on a new mode of food production that provided, for the first time in human history, sufficient food supplies to feed a larger number of people who did not produce food by themselves. Surplus production was turned from an occasional effort to a permanent institution. Four technological innovations yielded this change. The first was irrigation of arid land that would have been abandoned by horticulturalists when the soil was exhausted. If the desert or high mountains set limits to extending horticulture to new lands, food production can only be increased by multiplying the annual harvest or by turning the arid land into fertile grounds. This is exactly what happened in the river valleys of Nile, Euphrates, and Indus. Preserving the annual inundations of the river shores and controlling the flow of water by ditches and canals allowed to extend the fertile area and to have several harvests on a limited territory. For a long time irrigation systems were combined with advanced modes of horticulture.

Most important was, therefore, the change of tools for breaking and turning the soil. While horticulturalists used just a stick or a hoe for

B. Giesen et al., *Human Societies*, Cultural Sociology, https://doi.org/10.1007/978-3-032-07697-7_5

preparing the ground for the seed, the invention of the plough provided a way of turning the soil to a greater depth and to kill weeds that outgrow crops in simple horticulture after some time.

At first ploughs were driven by humans. After the domestication of cattle (around 6000 BC) and horses (around 3000 BC), however, human energy could be replaced by animal energy. For the first time in evolutionary history humans could not only harness energy from nature by nurturing, but they could use additional sources of energy. One person could plough a much larger piece of land per day than before and produce crops to feed not only his family but additional persons who were set free from food production. The surplus economy was born, distribution became an issue, centralized storage required protective walls, exchange got an economic meaning, full time specialists could sustain a living without producing food by themselves. Additional inventions like the wheel and animal driven carts facilitated and accelerated transportation even of heavy loads. Horse or camel riding expanded the range of mobility greatly and riding a horse became the distinctive feature of new social groups. At first horses were used with carts and chariots with spoked wheels (after 2000 BC). Cavalry came up not before 1000 BC. Up till the nineteenth century the material culture of rapid transportation in Europe, in Northern Africa, and in Asia was mainly based on horses.

The fourth technological breakthrough was the melting and forging of metal, first of copper (after 6000 BC) and bronze, later on of iron (after 1000 BC). It occurred also in several advanced horticultural societies before the invention of the plough. Metal tools are much more durable than wooden ones and lighter than stone tools, but metallurgy requires relatively sophisticated knowledge and manufacturing efforts. The command of metal manufacturing was, therefore, treated as a special talent and the use of metal was reserved for prestigious tools and items. A warrior equipped with metal weapons and armor in a horse driven chariot or, later on, on horseback was markedly superior to a traditional warrior who was protected only by leather and equipped with a stone axe. He was not only faster in attack and retreat, but also in striking and hitting and his weapons could easily pierce leather shielding. Bravery and heroism, which have been the decisive criteria for archaic warfare, could hardly compensate for this advantage in war technology. Thus society was divided into a smaller group of mobile warriors with metal weapons and horses and a large group of immobile peasants working the soil and producing food. Conquest and exploitation became possible. Warfare was now driven also by an economic motive.

In order to live on surplus produced by the peasants the warrior class needed a large population to exploit. Conquest societies are, therefore much larger than horticultural societies. Their population amounts hardly to less than several ten thousand and can consist of several million people.

Surplus production in ancient societies did not prevent miserable conditions for large parts of the population. Agriculture could produce more than needed for the survival of the producers, but this surplus was consumed by the small part of the population that was set free from the pressure to produce food. Starvation was therefore not a rare phenomenon in ancient conquest societies.

Classification: The Vertical Dimension of Social Life

Surplus production and metallurgy gave way to a new mode of social classification: social groups were not only thought as being separated by the boundaries of family and locality, but in addition to these horizontal distinctions, also by a vertical order that separated social class and rank. While the vast majority (up to 90%) of the population lived in villages as they did before, produced food, and imagined the world in familiar local contexts, a small group was seeing themselves to be positioned on top of these local villagers. Members of this group were set free from the pressure to produce food. Instead the adult male members of this group were mainly warriors who lived on the basis of tributes and enforced appropriation. These tributes were sometimes justified by pointing to the defensive tasks of the warrior class. But in most cases questions of legitimacy and justification did not come to the fore because the sheer threat of violence and brute force was sufficient to deter resistance and to ensure obedience and submission on the part of the peasant population. Whoever refused to pay tribute was killed or tortured. Ashurnasirpal the king of ancient Assyria commanded people who resisted to his rule to be skinned and the skins displayed on their houses.

> In pre-agrarian societies, devoid of stored wealth, no very clear sense can be attributed to power and domination. A man who does not provide for his own future, and who possesses no means for safeguarding the future, has only limited incentives for suborning others. Domination for immediate, non-delayed ends and for acquisition of status obviously can and does occur;

> but to what extent can it engender a permanent and impersonal stratification of power? The small scale of the society reinforces the barriers set up by the lack of permanent, storable resources.
>
> The situation changes drastically with the emergence of agriculture. Storable goods constitute a surplus. With the existence of stored value, power in a serious sense becomes not merely possible, but mandatory. Plato noted that it was the existence of a surplus which made defence and order-enforcement unavoidable. Basically, you own what you can defend, physically or socially. Property and power are correlative notions. You can defend property by direct force, or by invoking social rules which are themselves sanctioned by force. Domination in a propertyless society is not inconceivable. But there is clearly a limit to the persistence or elaborateness or scale of such domination. But there is no such limit, once material wealth makes its appearance.
>
> The agricultural revolution gave birth to extensive storage and wealth and thereby turned power into an unavoidable aspect of social life. No set of rules for the distribution of a surplus can be either self-evident or self-implementing. Mechanisms must exist for protecting resources, and for enforcing the "equitable" or approved distribution of their fruits. So power is born, and becomes the inescapable accompaniment of social order.
>
> Gellner, Ernest (1989): *Plough, Sword and Book. The Structure of Human History*. Chicago: University of Chicago Press, p. 145f.

The tribute system engendered new oppositional attitudes with respect to material production. Seen from the point of view of the warriors there was no limit to production and consumption: conquering new territories and new people allowed increasing personal wealth and military power. Since there were always competing conquerors outside, the conquest of new territories required larger armies than those of the enemies. Victories yielded additional resources of land and people and gave way also to new conquests, etc.

Seen from the point of view of peasants, by contrast, working more than necessary for survival yielded no returns. Therefore working was considered to be a burden and ways of avoiding work activities became a part of the social skills. Surveillance and punishment were the response of the warriors.

Military conquest and the brute violence marked the beginning of ancient class society, but the original reign of terror gave gradually way to an institutionalized boundary between a class that scorned working and lived on tribute and a working majority that had to produce this tribute and could not even imagine that this inequality could be changed.

The ruling class and the peasants were separated by a strong divide, intermarriage was almost impossible, sometimes both groups spoke different languages. In ancient Egypt the members of the ruling class considered the peasants to be closer to cattle than to themselves and the Roman writer Varro described slaves as "articulate goods" in distinction to cattle. Even in medieval England the children of peasants were called "litter." Economic inequality matched this social distance. In early modern England the household of the king consumed more than ten thousand times more money than the average household of a commoner.

Extreme social inequality is also reflected in the institution of slavery. Slavery turned defeated and conquered people into commodities that could be traded and used according to the wants of their owner. The sacred core of human existence was denied for them. Their fate varied largely, however. While, in the Roman Empire, galley slaves or mine slaves survived rarely the first year of their condition, house slaves were sometimes treated as members of the family and occasionally even released into freedom. In some ancient societies (e.g., Babylon) slaves could own property and acquire a certain wealth. Although slavery was common in ancient societies the slaves were never the majority of the population. In Roman Italy the estimate for the number of slaves is 30% of the total population. (Only in the early modern Caribbean system of plantation slavery the slaves were the majority of the population. Early Spanish colonialism established conditions of life for the indigenous population that bordered the worst of slavery in ancient societies.)

In general the institution of slavery was rarely questioned in ancient societies. The cultural basis for this (to our modern mind: amazing) lack of critical discourse about slavery is provided by the fundamental vertical distinction that was the cultural backbone of ancient societies: every person, every social group, every animal had its position in a great hierarchical order of being that extended between the sacred and the profane: Gods, heroes, warriors, peasants, slaves, horses, pigs, fishes, bugs, etc. If people in ancient societies wanted to understand something they had to define its position in the great hierarchy of being and in particular they had to relate this position to their own one. Profanity rules with respect to lower positions, awe, submission, and respect are due with respect to higher positions.

[...] one who is a human being belonging by nature not to himself but to another is by nature a slave, and a person is a human being belonging to another if being a man he is an article of property, and an article of property is an instrument for action separable from its owner. But we must next consider whether or not anyone exists who is by nature of this character, and whether it is advantageous and just for anyone to be a slave, or whether on the contrary all slavery is against nature. And it is not difficult either to discern the answer by theory or to learn it empirically. Authority and subordination are conditions not only inevitable but also expedient; in some cases things are marked out from the moment of birth to rule or to be ruled. And there are many varieties both of rulers and of subjects (and the higher the type of the subjects, the loftier is the nature of the authority exercised over them, for example to control a human being is a higher thing than to tame a wild beast; for the higher the type of the parties to the performance of a function, the higher is the function, and when one party rules and another is ruled, there is a function performed between them)—because in every composite thing, where a plurality of parts, whether continuous or discrete, is combined to make a single common whole, there is always found a ruling and a subject factor, and this characteristic of living things is present in them as an outcome of the whole of nature, since even in things that do not partake of life there is a ruling principle, as in the case of a musical scale. However, this matter perhaps belongs to an investigation lying somewhat outside our subject. But in the first place an animal consists of soul and body, of which the former is by nature the ruling and the latter the subject factor. And to discover what is natural we must study it preferably in things that are in a natural state, and not in specimens that are degenerate. Hence in studying man we must consider a man that is in the best possible condition in regard to both body and soul, and in him the principle stated will clearly appear—since in those that are bad or in a bad condition it might be thought that the body often rules the soul because of its vicious and unnatural condition. But to resume—it is in a living creature, as we say, that it is first possible to discern the rule both of master and of statesman: the soul rules the body with the sway of a master, the intelligence the appetites with constitutional or royal rule; and in these examples it is manifest that it is natural and expedient for the body to be governed by the soul and for the emotional part to be governed by the intellect, the part possessing reason, whereas for the two parties to be on an equal footing or in the contrary positions is harmful in all cases. Again, the same holds good between man and the other animals: tame animals are superior in their nature to wild animals, yet for all the former it is advantageous to be ruled by man, since this gives them security. Also, as between the sexes, the male is by nature superior and the female inferior, the male ruler and the female subject. And the same

> must also necessarily apply in the case of mankind generally; therefore all men that differ as widely as the soul does from the body and the human being from the lower animal (and this is the condition of those whose function is the use of the body and from whom this is the best that is forthcoming)—these are by nature slaves, for whom to be governed by this kind of authority is advantageous, inasmuch as it is advantageous to the subject things already mentioned. For he is by nature a slave who is capable of belonging to another (and that is why he does so belong), and who participates in reason so far as to apprehend it but not to possess it; for the animals other than man are subservient not to reason, by apprehending it, but to feelings.
>
> Aristotle (1967): *Politics.* London: William Heinemann; Cambridge: Harvard University Press, pp. 19–23

This culture of vertical inequality is mirrored by the distinction between master and servant that patterned social life in ancient societies. The master was defined by having servants and the servant was defined by having a master. Both positions constituted each other and both are dependent on each other. The master can be a master only insofar as the servant recognizes his master hood and this act of recognition conveys a certain sense of power to the servant. In distinction to the mass of peasants the servant lives in the household of the master and communicated with the master on a daily basis. This engenders a certain respect for the servant and sets the servant apart from the despised peasants.

Contrary to our modern mind, however, being a servant was not regarded as humiliating in a traditional class society. Instead the servant's position was a respected one—the servant belonged to a household of a master and was related to the great hierarchy of being.

In distinction to the servant the free person who neither belonged to a household nor owned one was frequently in a miserable situation. Free persons without a master and without servants did not fit into the social grid of traditional class society. For them, "freedom is just another word for nothing left to lose" as the famous rock song of Janis Joplin says. They had a scandalous and useless existence, were treated with contempt and disdain like dirt, and led a dangerous life as homeless vagabonds who had to rely on begging and stealing for survival. Identity was rooted in the land and in the hierarchy. Nomadism was thus shifted to the margin of ancient society. Nobody cared and complained if one of these vagabonds were killed or enslaved. This does not mean, however, that the number of vagabonding people without a master was small. In some ancient societies, more than a quarter of the total population was uprooted and on the move. Although despised and marginalized the vagabonding people were also regarded with a certain fear by the sedentary population. They were sometimes thought to command witchcraft and demonic powers and many people were afraid of their curse or bad spells. Therefore, even in later periods, beggars and gypsies are kept at a distance by giving them the little alimony they are asking for.

Cultural classification in ancient societies was not only patterned by the vertical division. There was also a new divide between the people living in the countryside on the one hand and the people who lived in the newly emerging towns and cities. The spread of cities in the near east started around 3000 BC. Like the villages these urban centers were permanent settlements but their population consisted mostly of merchants and craftsmen in addition to parts of the ruling class and their households. These merchants and craftsmen were fully specialized workers who exchanged their products against food, other products or precious metals.

In most ancient societies the urban population was less than 10% of the total population but it established a distinctive style of life that separated town and country. Both people of the city and people of the country were separated by a social distance: for the country people the city life appeared as decadent and cowardish, while the city people despised the country as uncivilized and brute.

In rural villages every person was familiar with the names and faces of the other villagers and strangers remained a rare and intimidating exception. The life in the city, however, demanded to encounter closely strangers whose name and face were not only unknown but remained that way. These strangers could neither be converted into familiar persons nor kept at a distance—the way rural villagers coped with them was blocked. Instead, their presence had to be disregarded. City people have to practice "civic inattention" to strangers who live or pass by at a close range.

> It is possible, then, to list some of the sources of social control in public places, but it is hard to have confidence that, even taken together, these are the relevantly effective ones. Whatever its causes, however, the surface character of public order can be identified. The inhibition of aggression routinely occurs, and some sort of civility usually is sustained. Individuals exert respectful care in regard to the setting and treat others present with civil inattention; mutual interference is avoided without much sign that an effort to do so is involved, and when accident or incident brings strangers into

entangling contact, brief remedial interchanges are extended. Expressions of civility presumably serve to assure recipients that they are the target of either no intentions or honorable ones, and this apparently reduces alarm even though actual vulnerability remains.

[…] The vulnerability of public life is what we are coming more and more to see, if only because we are becoming more aware of the areas and intricacies of mutual trust presupposed in public order. Certainly circumstances can arise which undermine the ease that individuals have within their *Umwelt.* Some of these circumstances are currently found in the semi-public places within slum housing developments and slum neighborhoods, and there is no intrinsic reason why some of these sources of alarm (as well as some additional ones) cannot come to be found in the residential community of the respectable classes, causing the fragile character of domestic settings to be evident there, too. Certainly the great public forums of our society, the downtown areas of our cities, can come to be uneasy places. Militantly sustained antagonisms between diffusely intermingled major population segments—young and old, male and female, white and black, impoverished and well-off—can cause those in public gatherings to distrust (and to fear they are distrusted by) the persons standing next to them. The forms of civil inattention, of persons circumspectly treating one another with polite and glancing concern while each goes about his own separate business, may be maintained, but behind these normal appearances individuals can come to be at the ready, poised to flee or to fight back if necessary. And in place of unconcern there can be alarm—until, that is, the streets are redefined as naturally precarious places, and a high level of risk becomes routine.

Goffman, Erving (1971): *Relations in Public. Microstudies of the Public Order.* New York: Basic Books, p. 331f.

The close proximity of strangers can no longer be treated as a violation of the boundary between inside and outside and it can no longer stimulate actions that produce distance and reconstruct the boundary. Instead the presence of strangers has to be accepted as a regular condition of interaction. Strangers are no longer unclassifiable subjects, but part of everyday life.

Although there are, of course, networks of neighborhood, professional association, or ethnic origin in ancient cities, the basic relationship toward the majority of other people in the city is impersonal and anonymous. Occasionally a merchant at a bazaar will engage into a short conversation with a possible customer in order to convey a sense of personal relationship, but this is just a pretext for business. Merchants exchange their wares not because they feel a personal obligation toward a customer or because they are linked to her by kinship ties, but instead because the offer is good. But there has to be a certain basis of trust even in these impersonal relations. The customer has to trust in the quality of the ware and the merchant has to trust in the value of the metal coins he gets in exchange. This indispensable trust in impersonal relations is mostly generated by controlling third parties—the professional association or guild on the one side and the ruler who issued the money on the other.

This impersonal coexistence of strangers applies, however, only for public spaces. In small villages or in simple societies the boundary between the "private" spaces of families and the "public area" was not strictly marked, people could observe the family life of neighbors, privacy was not yet shielded by fences and walls. Cities, by contrast, have to separate the public spaces clearly from the private households in order to cope with the presence of strangers. Access to the private realm is highly regulated and controlled by the master of the household, public spaces can be used by anybody. Here the temporal order replaces the social one: the principle "first come, first served" rules.

Separating the private from the public space is fostered also by a clearer demarcation of individual property rights. In ancient societies most rights were communal ones. The cattle or the house were owned by a clan, individual property rights extended mostly to weapons and decorative items. In ancient societies property rights on land became an important issue, houses had a personal owner, and wealth could be accumulated by exchange or forced appropriation. All these transactions required the institutionalization of individual property rights.

Circulation: Money, Tribute, and the Marcher Lords

Fulltime specialists in the crafts can operate on the basis of barter, that is, the exchange of material goods against other material goods or services. Barter requires, however, a "double coincidence of wants." This means, at first, that the craftsman can use the particular goods offered by the customer. This highly restrictive condition can be overcome if the needs of third parties are taken into account. The merchant may accept goods that he does not need but that can be used by many other people who will be ready to exchange them against something that the craftsman needs. This is the case with food, with grain, crops, or cattle. Animals have considerable advantage over crops: they are mobile and can circulate more easily than fruit or crops. Furthermore they will not decay as soon as crops, fruits, or vegetables. Hence the number of sheep, camels, or horses provided the first standard of payment in ancient societies. After the invention of metallurgy pieces of metal took the position of food. Everybody demanded it, it was durable and easy to circulate. These pieces of metal were standardized according to their weight and the weight was inscribed in it. After 1000 BC coins of copper, bronze, silver, and gold became thus the first primitive form of money.

Money has special social properties. It enables exchange beyond the special needs and wants of the seller and hence speeds up the circulation of goods and services. Unlike barter exchange the comparison of offers is easy and it can be used to deposit and store buying power. Thus economic power is disconnected from the moment when it is acquired: money can be accumulated and saved for the moment when it is needed.

Money depersonalizes social exchange: you accept an offer not because of personal obligation or kinship ties, but because the offer is the best one you can get. The person or the buyer does not matter and the history of the piece of money is of no concern: whether it had been stolen or used to pay a murder before, whether the money giver is trustworthy or not, whether the receiver will spend it for immoral purposes makes no difference. Money has no memory and interrupts social obligations. Neither the person who gave the money nor the one who received it have any obligations to each other after the exchange is completed. In this way money exchange is the opposite of gift exchange. It disregards social rank and local roots, its meaning is impersonal. Money represents nothing but general chances to exchange and buy.

> At the opposite pole from the mentality of the Middle Ages is the credit economy, in which the bill of exchange serves as money. The dominant idea in the Middle Ages was the substance of money, rather than its effects, which were in fact reduced to a minimum, though they could not really be eliminated altogether. In the modern conception of money based upon metal, the vital point is the working of the substance; and finally, in a credit economy there is a tendency to eliminate substance entirely, and to regard the effects as the only important matter.
>
> This superficial view owes something to an earlier scheme of thought which distinguished in all phenomena between substance and accident. Certainly this was historically of tremendous importance: the analysis of every phenomenon into a substantial core and relative, changeable

> appearances and qualities was a first orientation, a first guiding thread through the mysterious formlessness of things, giving them a structure and subjecting them to a general category that was consonant with our intellect. The mere sense differences that appear at first thus become organized and determined in their mutual relations. But it is the nature of such forms, as of social organizations, to appear and to claim to be eternal. Just as the destruction of one social order in favour of another gives the impression that all order and regularity is being abandoned, so does reconstruction in the intellectual realm. The objective stability as well as the subjective understanding of the world seems to be destroyed when a category that formed the backbone of a world view is abrogated. But money value will be just as little able to resist its reduction to functional value as light, heat and life can retain their particular substantial quality or resist their dissolution into forms of motion.
>
> Simmel, Georg (1990): *The Philosophy of Money*. London; New York: Routledge, p. 169f.

The power of money to circulation engenders a fluidification of social relationships. Stable ties between persons and objects tend to be dissolved and activities tend to be disconnected from non-monetary motives. However, all human societies have erected cultural barriers against a total profanization and commodification of social life. Some objects and actions are not for sale, they are exempted from money exchange. Cultures vary of course in the definition of what cannot be treated as a commodity. We, in contemporary Western societies, exclude human life and political office, the sexuality of children and the verdict in court from the realm of commodities, early modern European societies were particularly concerned with the venality of religious salvation, other cultures allowed for buying the sexual abuse of children, for trading slaves and for bribing judges because they had different definitions of the sacred. In some ancient societies (e.g., medieval Europe) the land was considered as largely exempted from money exchange because it bordered almost a sacred status. Land property was handed over from one owner to the other by inheritance or conquest or by feudal oath between seigneur and vassal. This rule ensured that the property on land remained largely within the hands of the warrior class.

Market exchange in the towns extended mostly to the products of local craftsmen or peasants. Only a few precious and rare items like metals, slaves, cloth, spices, or artwork were traded over a long distance by caravans and ships. Transportation over a long distance was cumbersome, dangerous, and slow. Its costs exceeded rapidly the value of the local substitute of the respective good. It made no economic sense to trade vegetables and crops from an area far away—they would perish before they could arrive.

Most economic transactions in conquest societies, however, were not mediated by money and markets. They were forced appropriations by the warrior class and its leader, the king. Tribute, taxes, and imposts funded the cost of living of the ruling class and the expenditures for military campaigns. Only at the beginning these forced transactions were seen as violent operations devoid of any legitimacy—in particular if several competing warrior gangs demanded tribute from the same people and if the tribute was strongly raised. When, after some time, conflicts between the competing warrior gangs were turned into a unified and centralized rule and the amount of tribute was standardized and expectable, the subjects were gradually accepting the tribute as a regular condition of life. The emergence of law based taxation is structurally similar to the transition from a mafia like system of irregular tributes to a state centered legal system of taxation. The first is based on the threat of direct violence, on physical presence of the warriors, and on oral communication, and the second is based on accepted claims, on an absent ruler, and on written law. Conquest societies struggled to achieve this transition and they rarely succeeded completely.

Most of them were able to overcome conflicts between different fractions of the warrior class and to establish a unified personal rule. Defeated enemies were turned into allies when they were faced with the choice of joining the victorious army or being enslaved. Thus the defeated enemy got "an offer that he could not reject" to use a phrase from *The Godfather*.

The authority of the ruler was based on his ability to keep peace inside and to defend the territory and the people against invasions and conquests from outside. Both goals could be achieved by the threat of violent intervention inside and of war outside. Furthermore war opened up chances to expand the territory by conquest. Therefore the ruling lords spent much of their tribute income for standing armies. Early Mesopotamian armies in the second millennium BC numbered up to 50,000 soldiers.

Successful conquerors had, however, to face a special risk of overstretching their conquered territory and of hence losing control. The maximum speed of a marching army is estimated at ca. 25 miles per day. Larger armies move much slower, the amount of provisions on food and water supplies needed sets limits to the mobility of armies. Given these limitations by speed and logistics the ruler and his army could reach the place of rebellion or the hostile invasion only with a considerable time lag. In large conquest empires like the one established by the Macedonian king Alexander it took half a year to travel from one end to the other. The charismatic presence of the ruler and the threat of his warriors ensured submission and peace, with growing distance, however, his authority was fading. Therefore the conqueror had to be constantly on the move to visit the different areas of his empire, to establish justice, to encourage his supporters, and to deter the enemies of his rule. We find this pattern of a moving center and of the unity of military, political and sometimes also religious leadership not only in ancient near eastern societies such as the Sumerian, Hethitian, Egyptian, Accadian, or Assyrian kingdoms, but also in ancient Greece, in medieval Europe, in Mongolian empires, in ancient China, in medieval India or in the Aztec kingdom.

> But where was the center? For a second time, the notion of fixed territories and cores gets in the way. For the center was the army, Sargon's [Akkadian King, B.G.] 5,400 men, and that was mobile. Only the ongoing campaign centralized military power. Once pacification and broader threats became uneven, the less the empire resembled an army engaged in a single campaign under its central leader. Provincial threats were responded to by mobilizing provincial armies, which put power in the hands of local commanders not the central state. To counteract fragmentation, the greatest conquerors in preindustrial communications conditions were in almost perpetual campaigning motion. Their physical presence in army headquarters centralized their power. Once they, or their successors, settled back into a court in a capital, the cracks usually showed. Indeed, many conquered empires then collapsed. We have not yet seen anything that might hold such artificial creations together except erratic fear and the energy of the ruler.
>
> One reason for instability was that no major advances in logistics had been made toward the *political* consolidation of empires. The state apparatus such as it was, depended on personal qualities and relationships of the ruler. Kinship was the most important source of permanent authority. But the greater the extent of conquest, the more kinship between ruling elites became strained and fictitious. In this period lieutenants intermarried with locals to secure themselves, but this weakened ties among the conquerors. In this period techniques of literacy were at first restricted to heavy tablets and complicated scripts.
>
> Mann, Michael: (1986): *The Sources of Social Power. Vol. I.* Melbourne: Press Syndicate of the University of Cambridge, p. 145

This rule of the marcher lords does not exclude that the ruler has homelands or fortified places where he spends some time of the year, but moving his presence through the realm is a central mode of establishing his authority in a larger territory. For the time of his absence he could establish a network of outposts or forts with some soldiers led by an appointed lieutenant. But this was hardly a substitute for the regular visit of the lord who as a conqueror owned the land and treated it as his personal property. This institutionalized rule of the land by the lord who had conquered it is called "patrimonialism." Patrimonial authority is according to Max Weber:

> Appropriation by the chief personally is a phenomenon of patrimonialism. It may vary enormously in degree to the extreme cases of a claim to full proprietorship of the land and to the status of master over subjects treated as slaves. Appropriation by particular social groups generally means the appropriation of at least a part of the means of administration by the members of the administrative staff. In the case of pure patrimonialism, there is complete separation of the functionary from the means of carrying out his function. But exactly the opposite is true of decentralized patrimonialism. The person exercising governing powers has personal control of the means of administration—if not all, at least of an important part of them. This was true of the feudal knight, who provided his own equipment, and of the count, who by virtue of holding his fief took the court fees and other perquisites for himself and met his obligations to his superior lord from his own means, in which these appropriated sources of income over which he had full control were included. Similarly, the Indian *jagirdar*, who provided and equipped a military unit from the proceeds of his tax benefices, was in complete possession of the means of administration. On the other hand, a colonel who recruited a mercenary regiment on his own account, but received certain payments from the royal exchequer and paid his deficit either by curtailing the service or from booty or requisitions, was only partly in possession of the means of administration and was subject to certain regulations. On the other hand, the Pharaoh, who organized armies of slaves or coloni, put his clients in command of them, and clothed, fed, and equipped them from his own storehouses, was acting as a patrimonial chief in full personal control of the means of administration. It is not always the formal mode of organization which is most decisive. The Mamelukes were formally slaves recruited by the purchases of their owner. In fact, however, they monopolized the powers of government as completely as any feudal class has ever monopolized fiefs.
>
> Weber, Max (1964): *The Theory of Social and Economic Organization*. New York: The Free Press, p. 349f.

An alternative way of coping with the large size of conquest societies leads to feudalism. Here the ruler lends parts of the conquered territories to warriors in exchange to a sacred oath to support the ruler in military conflicts. By the feudal contract between seigneur and vassal the conqueror is discharged to control the land and its taxation and to fund a large standing army of warriors. Substituting the ruler's presence, his vassals remain in the area, exploit the land by taxation and, in return, have to deliver military support to the ruler.

The feudal solution of the control problem has, however, some serious disadvantages for political authority. It decentralizes and diffuses political authority significantly. The vassals do not only face the seigneur on an equal footing in the feudal contract, but they also control actually the land and dispose of the military force. After a century of feudalism the power of the Carolingian emperor in early medieval Europe was much smaller than the power of his mighty vassals. Similar developments can be found in early modern Japan or in ancient China in the period of the warring lords. The vassals on their turn can engage in wars against each other and the ruler has hardly enough own warriors to reestablish peace in his realm. Thus feudalism engenders frequently in a continuous internal war and a diffusion of central political authority.

Political authority in ancient societies was not based only on the fear of violence and terror alone. Performing terror and executing violence is a very costly way of political control. Therefore, after the conquest, the rulers try to replace it by symbolic means. The authority of the ruler is to be accepted by the subjects because they believe it to be legitimate. In ancient societies the main reason to believe in this legitimacy is tradition and descent.

> A system of imperative co-ordination will be called 'traditional' if legitimacy is claimed for it and believed in on the basis of the sanctity of the order and the attendant powers of control as they have been handed down from the past, 'have always existed.' The person or persons exercising authority are designated according to traditionally transmitted rules. The object of obedience is the personal authority of the individual which he enjoys by virtue of his traditional status. The organized group exercising authority is, in the simplest case, primarily based on relations of personal loyalty, cultivated through a common process of education. The person exercising authority is not a 'superior', but a personal 'chief'.
>
> His administrative staff does not consist primarily of officials, but of personal retainers. Those subject to authority are not 'members' of an

> association, but are either his traditional 'comrades' or his 'subjects.' What determines the relations of the administrative staff to the chief is not the impersonal obligation of office, but personal loyalty to the chief.
>
> Obedience is not owed to enacted rules, but to the person who occupies a position of authority by tradition or who has been chosen for such a position on a traditional basis. His commands are legitimized in one of two ways: (a) partly in terms of traditions which themselves directly determine the content of the command and the objects and extent of authority. In so far as this is true, to overstep the traditional limitations would endanger his traditional status by undermining acceptance of his legitimacy. (b) In part, it is a matter of the chief's free personal decision, in that tradition leaves a certain sphere open for this. This sphere of traditional prerogative rests primarily on the fact that the obligations of obedience on the basis of personal loyalty are essentially unlimited. There is thus a double sphere: on the one hand, of action which is bound to specific tradition; on the other hand, of that which is free of any specific rules.
>
> In the latter sphere, the chief is free to confer 'grace' on the basis of his personal pleasure or displeasure, his personal likes and dislikes, quite arbitrarily, particularly in return for gifts which often become a source of regular income. So far as his action follows principles at all, these are principles of substantive ethical common sense, of justice, or of utilitarian expediency. They are not, however, as in the case of legal authority, formal principles. The exercise of authority is normally oriented to the question of what the chief and his administrative staff will normally permit, in view of the traditional obedience of the subjects and what will or will not arouse their resistance. When resistance occurs, it is directed against the person of the chief or of a member of his staff. The accusation is that he has failed to observe the traditional limits of his authority. Opposition is not directed against the system as such.
>
> Weber, Max (1964): *The Theory of Social and Economic Organization.* New York: The Free Press, p. 341f.

The ruler is presented as the son of the previous ruler who, on his turn, is presented as the son of the previous ruler, etc. At the top of the social hierarchy the claim of authority is justified by translating the well-known reference to ancestors into a detailed account of descent. Political legitimacy relies on sacredness of ancestors and the continuity of hierarchical order.

Lineage and descent are, however, not obvious and visible for the subjects. Therefore the history of the dynasty has to be represented, imagined, and narrated, the ruler has to use the traditional signs and symbols of his dynasty in public, his name has to indicate his lineage, etc. As the ruling lord embodies the unity of the society the dynasty defines and delimits the continuity of history. It is not the continuity of an ethnic group, but the dynastic continuity of authority that provides the persisting subject of history. In the context of ancient societies innovation and novelty appear as immoral attacks on the sacred tradition. The only way of legitimate change consists in restoring the origin and returning to the good old traditions that are put at stake by innovations. But it was not only traditional legitimacy that disguised the cruelty of the original conquest. Its violence and terror also have to be covered by mythical stories about the heroic foundation of the state: defeating demons and monsters, migrating into the Promised Land, golden ages of prosperity and power, etc.

Representation: Monuments and Processions, Temples, and Texts

After 3000 BC cities and stone architecture gave way to new modes of representation. Visual representations can now be blown up to immense sizes and their stone materials endure the wear of time. Thus the mortal person of the king can overcome the restrictions of time and space. By symbolic representation he can counteract his mortality and the limitations of his bodily presence. The dead ruler can be represented by huge monuments and images that depict him as the founding hero of the state, glorify his victories, and tell his life to further generations. The heroic conqueror is imagined as exceptional and superhuman, he defies reason and danger, ventures out into the unknown, defeats monsters and enemies before he returns with glory and riches to his home. By telling his heroic story in stone representations, in pyramids, monuments, and reliefs, the ruler partakes in the eternity of gods.

But it is not only the mortality of the ruler that is overcome by images and monuments. We have seen above that the presence of the ruler is of paramount importance for his authority. His unavoidable absence in most places of the realm can be compensated by statues and images. These images of the ruler differ from relics, but they are also not yet just exchangeable representations that are clearly distinguished from their reference. They are symbols partaking in the sacred nature of the represented person. Denying respect to them amounts to an act of profanization.

The same holds true for the statues of gods in temples. They too are symbolic embodiments of the sacred, they are fused with sacrality and whoever is close to them partakes in their radiating aura. The gods are eternal, they can appear in many forms, but they are not everywhere. Therefore it makes sense to visit their places, to sacrifice, and to pray there. Every city and every lordship had their own tutelage God who was venerated in the main temple, but this temple could also accommodate statues of other deities, some of them tutelage gods of other cities. Even if the tutelage God got an exclusive temple there were other temples for other gods in the city. The polytheistic world of ancient conquest societies was quite tolerant in cultural matters.

This religious tolerance and coexistence is due to a special relationship between the sacred and the profane. Both spheres, the sacred and the profane, the visible world of objects, and the invisible world of agency and subjectivity are not yet strictly separated. The sacred is unevenly distributed in reality and there are transitions between the sacred and the profane—heroes who are half human and half divine and who can work miracles, gods who are moved by human passions, who walk among the humans and talk to them, double creatures who are half human and half animal, etc. Here, too, the great chain of being creates many transitions and mediations between above and below, heaven and hell, the sacred and the profane. Since the sacred is embodied in persons, objects, and places it can coexist with other embodiments of the sacred.

But the logic of symbolic representation is not limited to the imagination of gods and lords as important and central embodiments of society. Stone buildings could also symbolize the basic structure of ancient society. Let us just mention some features of ancient architecture: stairs, columns, and courts. Ancient temples and palaces abound with stairs. These stairs are not simply material devices to connect different physical levels for walking persons. Instead, they represent the vertical dimension of social classification; they uplift the ruler or the priest on top of the stair and humiliate the person below, thus compensating for the undeniable fact that rulers and priests are, in principle, not taller or higher positioned than regular citizens. At the same time stairs represent the many transitions between high and low, the graded nature of social stratification.

> Given the symbolic importance of highness, many communities celebrate their corporate authority by investing it in the physically most striking symbols. The tallest structure of a community often becomes its representative

> feature: among cities it is a particular skyscraper; in small towns and villages the spire of the church is the predominant collective symbol. In both types of community the most imposing vertical structures are built by the most central social institutions. This tendency has maintained itself well over time. The skylines of medieval cities are dominated by religious institutions; modern cities, by economic institutions; capital cities, in both eras, by political institutions. The contrasts thus produced are often sanctioned by ordinances which prohibit the erection of any structure taller than a specified sacred edifice [...], ensuring that the most representative edifice will be vertically prominent. Given the sacred qualities of these high structures, tourist pilgrimages are often made to them. On the other hand, the height of a structure may be used to symbolize the power of its residents.
>
> Schwartz, Barry (1981): *Vertical Classification. A Study in Structuralism and the Sociology of Knowledge*. Chicago: University of Chicago Press, p. 62

A similar symbolic meaning is carried by pyramids in ancient cultures. They do not serve practical or profane purposes—despite all our efforts to impute such practical purposes to them. Instead they link the vertical dimension of the great chain of beings to the idea of a unified center of society which was the godlike ruler, the Pharaoh, in ancient Egypt or to the central deities, for example, in Teotihuacan or Tenochtitlan in ancient Mesoamerica. Columns, too, have a symbolic meaning: they connect the soil with the roof and mediate thus between the heaven and the earth, between high and low, the sacred and the profane.

Complex buildings like temples and palaces do symbolize the central axis of classification not only by the extension of height but also by walls and enclosures. Again: the manifold forecourts and inner courts, anterooms and siderooms, gates and entrances are not to respond to practical exigencies but to symbolize multilayered boundaries and to shield the Arcanum of palace and temple from the eyes of ordinary people. Access to the sacred center of political authority or religious cult was possible only for those whose dignity was certified. In ancient societies the sacred is still embodied in objects and persons and since it is embodied it is subject to a special dialectics of veiling and revelation. If the sacred would be a visible part of everyday life it would risk to lose its extraordinariness. In order to avoid profanization it has to be covered and veiled most of the time. But since the embodied sacred works only through its presence, by contagion at close distance, it has to be displayed in public from time to time. On these occasions the public space and the people will be fused by the radiation of the sacred objects or the aura of the ruler. The charismatic ruler or the embodiment of the sacred leave the Arcanum of the palace or the temple and appear in a solemn public procession. Strict ceremonial rituals regulate this close encounter with the people. The ruler rides a special animal, is accompanied by guardsmen and courtiers, blesses the people who hail him and try to come close to him. This is the charismatic moment of effervescence, in which the sacred as embodied in the ruler is represented ritually.

Rituals represent also the submission and subjugation of defeated enemies. Faced to the victorious lord and his warriors the defeated persons have to kneel down or to prostrate themselves on the soil and to beg for mercy. They are barefooted and sometimes naked to indicate their status outside of the community of respected persons. In many cases these rituals of humiliation and surrender end up with the lord generously granting the life of the submissive person, but he does not have to do so. Anyway the lord is presenting himself as the source of life and death.

The second new medium of symbolic representation in ancient societies is writing. The first known cases of writing are clay tables from the Mesopotamian city of Uruk dated around 3300. The invention of writing seems to parallel the rise of the cities.

> From that time, the capitals of the two adjacent regions played a decisive part: Uruk in Sumer and Susa at the foot of the Zagros mountains which saw the emergence of the first states worthy of the name, firstly with a break from the prehistoric tradition symbolised by painted pottery, then by the elaboration of a system of accountancy which had become essential to the management of their vast riches. This book-keeping led naturally to the creation of a system of writing, still primitive, partially pictographic but mainly abstract, which later became known as cuneiform and was adopted and adapted by the greater part of the Ancient East. This writing is found in Uruk at the end of the period of that name, around 3300 BC when their neighbours, of the same culture, only practised numerical accounting. Writing and book-keeping were established by a priestly administration that patronised an art resolutely realist in opposition to the stylized works of the prehistoric period.
>
> Amiet, Pierre (1982): "Introduction historique", cited in Goody, Jack (1986): *The Logic of Writing and the Organization of Society*. Cambridge: Cambridge University Press, p. 49

Writing is, of course, one of the greatest breakthroughs in symbolic representation. It allows to diffuse and reproduce information even when the person who wrote it is absent or already dead and the accuracy of written communication is much greater than oral reproduction. Furthermore writing allows accumulating knowledge beyond the capacities of the individual memory. Thus knowledge that would have been forgotten in oral cultures can be stored and held available for later users.

The core domain of texts is—apart from accounting—narrating stories and commanding actions and it is this domain to which the first uses of writing in history applied. The writing in ancient societies was mainly geared to tell the life of great rulers, to fix laws and contracts set by these rulers, to standardize measures and taxes and to address the subjects by letters. The first known codes of law were issued by Mesopotamian kings like Lipit-Ishtar and Hammurabi at the beginning of the second millennium BC.

Like oral language writing opens up ways to modify the represented reality: it can command, negate, forbade, question, etc. Therefore it is much better able to represent normative expectations or privileges than pictorial or performative media do. The rise of the legal institutions is, hence, closely linked to the invention of writing.

In ancient societies writing was a skillful art that was professionally performed by special scribes. These scribes worked mainly for the ruling lord. For common people writing as well as reading appeared to be an exclusive privilege of scribes not only because the material carriers of writing were expensive but also because of the long-term training required.

At the end we should note that writing, like all media, can represent reality only in a restricted way. Images, statues, or buildings can render much more accurately proportions, textures, and colors than writing while these media cannot tell stories or represent change.

References

Aristotle (1967): *Politics*. London: William Heinemann; Cambridge: Harvard University Press.

Childe, Gordon (1951): *Social Evolution*. New York: H. Schuman.

Gellner, Ernest (1989): *Plough, Sword and Book. The Structure of Human History*. Chicago: University of Chicago Press.

Goffman, Erving (1963): *Behavior in Public Places. Notes on the Social Organization of Gatherings*. New York: Free Press of Glencoe.

Goffman, Erving (1971): *Relations in Public. Microstudies of the Public Order*. New York: Basic Books.

Goody, Jack (1986): *The Logic of Writing and the Organization of Society*. Cambridge: Cambridge University Press.
Lenski, Gerhard E. and Jean Lenski (1974): *Human Societies. An Introduction to Macrosociology*. New York: McGraw-Hill.
Lovejoy, Arthur Oncken (1966): *The Great Chain of Being. A Study of the History of an Idea*. Cambridge: Harvard University Press.
Mann, Michael (1986): *The Sources of Social Power. Vol. I.* Melbourne: Press Syndicate of the University of Cambridge.
Schwartz, Barry (1981): *Vertical Classification. A Study in Structuralism and the Sociology of Knowledge*. Chicago: University of Chicago Press.
Simmel, Georg (1990): *The Philosophy of Money*. London; New York: Routledge.
Weber, Max (1964): *The Theory of Social and Economic Organization*. New York: The Free Press.

Empires

Bernhard Giesen

Production: Roads and Fortifications

Cities in conquest societies could be quite large. Sometimes their population exceeds 50,000. In order to feed cities of more than 100,000 inhabitants, however, the transportation system of conquest societies was too deficient. Long distance transportation was limited to precious goods that did not perish during the long time when they were moved by caravans.

Empires are based on rapid ship transportation on rivers and oceans and on an elaborated system of roads overland. Ship transportation existed also before the rise of empires but it got a new dimension with the invention of large ships and new techniques like rowing and sailing. The Chinese Junks, the Roman Galley, the Arab Dhau, or the Spanish Caravel provided the central mode of expanding the respective empires. The Chinese excelled in river transportation and even constructed canals to shorten connections. The Roman Empire disposed not only of rapid ship transportation on the Mediterranean sea that worked as an internal ocean of the empire, but had also constructed a relatively dense network of all-weather roads that connected the capital city with the provinces and that persisted even after the decline of the empire. The Inca Empire is another example of admirable road construction—in this case even without using carts and wheels. By these roads long-distance communication and circulation could be accelerated significantly, trade could extend also to goods that would have perished if transported for months. The constructive activity

B. Giesen et al., *Human Societies*, Cultural Sociology,
https://doi.org/10.1007/978-3-032-07697-7_6

of the Roman Empire was, however, not limited to roadwork and bridges. It produced also long-distance pipelines that supplied water for arid regions and expanded the reach of irrigation.

Thus capital cities like Rome, Alexandria, Byzantium, Baghdad, Hang Chou, or Kai Fang received supplies from distant parts of the respective empires to support their population which numbered sometimes more than a million persons. The total population of empires could exceed easily 50 million. The Roman Empire comprised more than a quarter of the world's total population at its time, the Chinese Empire counted more than 110 million people at the end of the eighteenth century, etc.

Most important were the consequences of road transportation for the system of military domination. Armies could be moved relatively quickly from one region of the empire to another in which they were needed. Hence the total number of soldiers could be reduced or the degree of control could be increased. Actually the number of soldiers in the standing armies of empires was quite high. An estimated 13% of the total population of the Roman Empire was in the military service.

The system of marcher lords was substituted by an institutional separation between the ruling emperor who lived in the capital city and the military leaders who stayed with their armies at the frontier. This separation of the military and the political leadership turned the emperor from a warrior to the embodied symbol of the unity of the empire and to the center of court life. It engendered also the risk of conflicts between military leaders and political leaders who had to be protected by special guards against attacks by unruly generals or politicians who occasionally tried to seize power by killing the emperor.

Among the most striking constructions in ancient empires are the fortifications at the frontier to defend the internal space of the empire against the raids of "barbarian" tribesmen. The great wall of northern China and the Roman Limes in Central Europe are the most famous and spectacular examples of these fortifications. Empires are defined by a moving frontier that separates their internal realm of peace from the world outside that was regarded as wild, uncivilized, and raw. Extending this internal realm of civilization and peace by war at the frontiers was seen as the self-imposed mission of the empire. Since the counterattacks of the tribesmen were rarely large battles, but, instead, unexpected raids, it was difficult to cope with them by the traditional warfare of standing armies. Fortifications and walls could reduce the risk of sudden raids and keep the number of soldiers defending the empire at the frontier relatively low.

Classification: Center and Periphery, the Rise of Officials, and the Axial Age Divide

Empires add to the vertical structures of conquest societies a new distinction: the distinction between center and periphery. Empires are centered on the person of the emperor, on a capital city, on a dominant ethnic group and a dominant language, and their administration and economy were centralized. In distinction to conquest societies with their marcher lords and moving centers empires are much more territorial. They have a stable center, a pacified inner realm and a periphery. This periphery consists of a diffuse and variable frontier zone that marks not only the boundary of the polity, but also the transition between civilization and wilderness, peace and war, culture and nature. Empires can hardly imagine a legitimate political entity outside of this frontier zone. They consider the outside to be clearly inferior and lacking "true" identity. But it is not only despise and derogation that is typical for the imperial attitude toward the outside. Empires are also driven by the missionary zeal to turn the barbarian outside into a civilized inside and to expand the frontier to include the inferior barbarians who do not yet know about their "true interests" to become subjects of the empire. Sometimes the frontier was protected not just by a walled fortification or a range of military forts but by a protective belt of allied barbarians with whom the empire had friendly relations and who received, in exchange, money or goods.

As intolerant to the outside beyond its frontiers empires are, as tolerant as many empires with respect to ethnic diversity inside. Because empires include quite diverse ethnic groups they have to avoid excluding them strictly and radically from the center. Empires are, by constitution, multicultural and represent sometimes the conquered groups in the titles of the emperor: the Roman emperor is also titled, for example, King of Cartage, Prince of Judea, Pharaoh of Egypt, etc. The elites of conquered ethnic groups are integrated by education and marriage, the other members of these groups are frequently granted citizenship rights. Integration by intermarriage or citizenship was, however, limited in the case of modern colonial empires. Here education was granted, but intermarriage between the colonizers and the colonized was hardly accepted.

This integration presupposes, however, that there is a central culture and language in which, by principle, every citizen can participate. Latin in ancient Rome, Mandarin in ancient China, Arabic in the Muslim empires, Spanish in the early modern Spanish colonial empire, or English in the modern British Empire were such central languages that were usually associated with a certain lifestyle and culture. The pervasive influence of this imperial culture was sometimes so strong that it survived even military defeat. After the destruction of the Roman Empire by northern tribes in the fifth century AD the Latin culture persisted and provided the medium of communication for European elites during the medieval times. Up till the nineteenth century no European could claim to be civilized if she was not able to read Latin. The Chinese civilization, too, survived several conquests by barbarian tribes from the north. As devastating as these conquests have been they also proved the superiority of the Chinese civilization. After a few generations the ruling Mongol or Manchu elites were sinified. Former British colonies, too, kept English as their dominant language after independence, etc. This superiority of language and civilization is created and carried by the structure of imperial politics but it persists after the collapse of the empire as a political power. One of the main reasons for this enduring and persisting existence of dominant languages is, of course, writing and reading. The dominant language in an empire is the language of administration, law, and literature whereas the languages of the conquered groups frequently exist only in vernacular form. When the conquered ethnic group possessed of a written language of their own this language has much greater chances to persist on their turn.

The literary language of the center, however, could not replace completely the original vernacular language of the ethnic communities. Therefore citizens of empires are usually bilingual and have a hyphenized identity: they are Syrians and Roman citizens, Jews and subjects of the Caliph in Baghdad, Indians and British citizens, etc.

In a similar way the places of residence in the large cities reflect the ethnic ties of those who have moved from the provinces to the big cities: there was a Greek quarter in ancient Rome, there was a Jewish quarter in Byzantium etc. In the big capital cities many immigrant communities complete the original native population. Thus the capital city mirrors the multiculturalism of empires.

Uprooted people who were outsiders in conquest societies because they could not be fitted into its ranking system got a different position in the capital cities of the great empires. Here masses of them lived close together in public spaces as beggars and prostitutes, day laborers, or street merchants. In some cities these urban poor amounted to more than a quarter of the total population. This opened up new chances for collective action. They could band together and, stimulated by a leader, they could rally and riot against the representatives of public order. This massed challenge to public order was difficult to meet in large urban environments. In distinction to slaves these people were free citizens who mattered as political supporters. Sometimes they were clients of influential citizens who acted as their patrons in political arenas, sometimes it was just their massed voices in public spaces that conveyed political weight to them. Therefore the proletarian masses of the capital cities had to be appeased by "bread and games" as the Roman phrasing goes. Otherwise rebellion and pillaging was imminent. These uprisings of the lower classes could hardly topple the system of domination but sometimes it took considerable military efforts to overcome them. Thus the uprooted masses were turned from outsiders of rank society to a political power in the empires.

Some marcher lords of conquest societies had already codified rights and issued laws thus separating laws from oral commands. But the gradually emerging sphere of law was mostly restricted to ad hoc jurisdiction by the ruler and it not yet administered by trained specialists. In empires, by contrast, the rule by family ties or by confidents is replaced by the rule of *legal specialists* who were not only able to read and write, but were also experts in applying the law and in systemic codification of the law. These legalist experts had got their education in special schools and centers of scholarship and had to undergo special examinations before they were appointed to an office by the emperor or one of his representatives.

The emperor established an extensive network of these offices whose holders were authorized to act as his representatives in public matters and who were remunerated for their activities by him. They could administer justice and taxes, supervise public works and decide in his name, command troops and grant privileges. Thus the absence of the ruler was to be compensated by official representatives who were equipped with his authority.

The legalist rule by office was based on two innovations: the first one was the transition from law as privilege to law as a written system of principles and the second was the institutionalized and sometimes even embodied disregard of family ties. If an official has a family to care for this care will prevail over the exigencies of the office. In particular if the person of the ruler is absent and tight controls are difficult, the institutionalized disregard of family ties is of crucial importance for the power of the ruler. While landowners can pass their lands to their offspring this possibility is denied to clerics and officeholders. Their power is a borrowed one. Thus it is not by coincidence that the main carriers of medieval legalism were monks. In some empires the officeholders were not denied to have a family but the family had to live in a different province. In other cases the officeholders were even bodily castrated. Occasionally this cutting of family ties was reinforced by using marginalized ethnic groups like the Janissaries, that is, Christian slaves, as military elites in the Ottoman Empire or foreigners like the Jesuits in early modern China. The most impressive case is certainly that of the Eunuchs at the Chinese imperial court.

More important than the Eunuchs were, however, the Mandarins in the Chinese Empire. Since the Song dynasty at the latest they replaced the old hereditary aristocracy as the main carrier group of the imperial rule. The Mandarins were scholar officials who reached their positions in the imperial bureaucracy and in civil service only by passing a highly competitive examination following a long and arduous study of the classical scriptures. Thus they got their position neither by conquest nor by descent or wealth, but by education, examination, and appointment. This new impersonal principle of access and the special training in impersonal justification patterned also the ways of thinking in this class of officeholders, clerics, and administrators. They conceived of the law not in terms of single privileges and traditions, but, instead, in terms of general principles and impersonal rationality. The new legalist officeholders tended to oppose the diversity of local customs and personal privileges, they underlined the principle and disregarded the exception, and they were moved by the missionary thrust to reorganize society on the basis of written law and principled rationality.

Writing and principled law fostered the idea of society as an abstract and invisible entity separated from the visible and tangible world of everyday life where people had bodies, faces, and names. In distinction to oral communication, written texts allow for disembodied and decontextualized communication. We can write a text without knowing its future readers and we can read a text without imagining its author or the situation in which the text was written. (In a similar way property rights once instituted can be transferred from one person to another and the new owner does not have to know the previous ones to enjoy the property right. Thus writing and legal rights open up, in principle at least, new spheres of circulation.)

This divide between the visible and tangible world of everyday life and the invisible and impersonal world of abstract principles fostered also the emergence of a new relationship between the sacred and the profane. Until the advent of the so-called axial age breakthroughs after 800 BC the sacred could be embodied in objects, persons, and places. It was unevenly distributed in the visible world but it could be part of this world and its agents, the gods, were imagined as behaving very similar to human beings. They could be jealous and generous, they could respond to sacrifices by grace and benevolence, and above all they could suspend the regular profane course of things: they could work miracles. As the gods had a human face, the kings were divinized. They were thought to be extraordinary and heroic, exempted from the profane limitations of everyday life. This relationship between humans and deities changed profoundly when the threshold to the so-called axial age was passed.

> What is new about this age, in all three areas of the world, is that man becomes conscious of Being as a whole, of himself and his limitations. He experiences the terror of the world and his own powerlessness. He asks radical questions. Face to face with the void he strives for liberation and redemption. By consciously recognising his limits he sets himself the highest goals. He experiences absoluteness in the depths of selfhood and in the lucidity of transcendence.
>
> All this took place in reflection. Consciousness became once more conscious of itself, thinking became its own object. Spiritual conflicts arose, accompanied by attempts to convince others through the communication of thoughts, reasons and experiences. The most contradictory possibilities were essayed. Discussion, the formation of parties and the division of the spiritual realm into opposites which nonetheless remained related to one another, created unrest and movement to the very brink of spiritual chaos.
>
> In this age were born the fundamental categories within which we still think today, and the beginnings of the world religions, by which human beings still live, were created. The step into universality was taken in every sense.
>
> As a result of this process, hitherto unconsciously accepted ideas, customs and conditions were subjected to examination, questioned and liquidated. Everything was swept into the vortex. In so far as the traditional substance still possessed vitality and reality, its manifestations were clarified and thereby transmuted.
>
> The *Mythical Age*, with its tranquillity and self-evidence, was at an end. The Greek, Indian and Chinese philosophers were unmythical in their decisive insights, as were the prophets in their ideas of God. Rationality and rationally clarified experience launched a struggle against the myth (*logos* against *mythos*); a further struggle developed for the transcendence of the One God against non-existent demons, and finally an ethical rebellion took place against the unreal figures of the gods. Religion was rendered ethical, and the majesty of the deity thereby increased. The myth, on the other

hand, became the material of a language which expressed by it something very different from what it had originally signified: it was turned into parable. Myths were remoulded, were understood at a new depth during this transition, which was myth-creating after a new fashion, at the very moment when the myth as a whole was destroyed. The old mythical world slowly sank into oblivion, but remained as a background to the whole through the continued belief of the mass of the people (and was subsequently able to gain the upper hand over wide areas).

This overall modification of humanity may be termed *spiritualisation*. The unquestioned grasp on life is loosened, the calm of polarities becomes the disquiet of opposites and antinomies. Man is no longer enclosed within himself. He becomes uncertain of himself and thereby open to new and boundless possibilities. He can hear and understand what no one had hitherto asked or proclaimed. The unheard-of becomes manifest. Together with his world and his own self, Being becomes sensible to man, but not with finality: the question remains.

For the first time *philosophers* appeared. Human beings dared to rely on themselves as individuals. Hermits and wandering thinkers in China, ascetics in India, philosophers in Greece and prophets in Israel all belong together, however much they may differ from each other in their beliefs, the contents of their thought, and their inner dispositions. Man proved capable of contrasting himself inwardly with the entire universe. He discovered within himself the origin from which to raise himself above his own self and the world.

In *speculative thought* he lifts himself up towards Being itself, which is apprehended without duality in the disappearance of subject and object, in the coincidence of opposites. That which is experienced in the loftiest flights of the spirit as a coming-to-oneself within Being, or as *unio mystica*, as becoming one with the Godhead, or as becoming a tool for the will of God is expressed in an ambiguous and easily misunderstood form in objectifying speculative thought.

It is the *specifically human in man* which, bound to and concealed within the body, fettered by instincts and only dimly aware of himself, longs for liberation and redemption and is able to attain to them already in this world—in soaring toward the idea, in the resignation of ataraxia, in the absorption of meditation, in the knowledge of his self and the world as *atman*, in the experience of *nirvana*, in concord with the *tao*, or in surrender to the will of God. These paths are widely divergent in their conviction and dogma, but common to all of them is man's reaching out beyond himself by growing aware of himself within the whole of Being and the fact that he can tread them only as an individual on his own. He may renounce all worldly goods, may withdraw into the desert, into the forest or the mountains, may discover as a hermit the creative power of solitude, and may then return into the world as the possessor of knowledge, as a sage or as a prophet. What was later called reason and personality was revealed for the first time during the Axial Period.

What the individual achieves is by no means passed on to all. The gap between the peaks of human potentiality and the crowd became exceptionally great at that time. Nonetheless, what the individual becomes indirectly changes all. The whole of humanity took a forward leap.

Jaspers, Karl (1965): *The Origin and Goal of History*. New Haven; London: Yale University Press, pp. 2–4

The first axial age breakthroughs that occurred between 800 and 400 BC in Israel and Iran, Greece and India replaced the realm of interaction between gods and humans by a strong tension between the profane this-worldly sphere on the one hand and the other-worldly and transcendental sphere on the other hand. The profane sphere, the visible world at hand was seen as profoundly and unredeemably unholy and strictly separated from the other-worldly and transcendental sphere. This one was no longer populated by a variety of gods who coexisted and had conflict prone relationships. Instead it was governed by unified impersonal principles exempted from the confines of time and space. Some visions of this transcendental sphere did not permit personal representations of the sacred at all, others imagined one almighty God who did not allow any other gods besides him. Monotheism tends to engender intolerance.

Trying to influence God's decisions, to change his laws or to change the impersonal principles of the cosmos were seen as impossible and even as blasphemous. No miracles and magic, no mediation any more. Polytheism, sacrifices, and sacred statues were denounced as pagan idolatry. Some of these transcendental visions went as far as banning any images of the sacred as blasphemy. The statues and images of the old polytheistic cults were frequently attacked and destroyed. The first case of this militant iconoclastic monotheism was pushed forward by the Egyptian Pharaoh Akhenaton who banned the ancient polytheistic cults during his rule.

In contrast to the original polytheistic religions that were based on sacrifice and miracle the new transcendental visions were missionary and intolerant: they claimed that there is only one God and one truth. Whoever resisted to conversion and rejected the message of salvation was despised as inferior nonbeliever and even seen as a dangerous being that did not deserve to live.

The new transcendental religions aimed at universal inclusion and in doing so their dynamics were similar to the dynamics of empires. Their strict insistence on the tension between the profane world and the transcendental sphere impeded, however, any effortless blending of the politics of empires and the missionary zeal of axial age religions. Politics and power as well as money and economics were regarded as utterly profane and there was no way of accommodating with the exigencies of this-worldly life. There were only two ways to cope with the tension between transcendental principles and profane temptation and thus achieve salvation from this world: the first consists of retreat from profane ambition and ascetic withdrawal from the pleasures of life, the second recommends action to change the profane and sinful earthly life and to realize the transcendental principles or the law of God within this world. Buddhism centers the first way to salvation, Protestantism and Islam the second one.

> We have had repeatedly to use the terms 'asceticism' and 'mysticism' as polar concepts. In order to elucidate this terminology we shall here further differentiate these terms.
>
> In our introductory comments we contrasted, as abnegations of the world, the active asceticism that is a God-willed *action* of the devout who are God's tools, and, on the other hand, the contemplative *possession* of the holy, as found in mysticism. Mysticism intends a state of 'possession,' not action, and the individual is not a tool but a 'vessel' of the divine. Action in the world must thus appear as endangering the absolutely irrational and other-worldly religious state. Active asceticism operates within the world; rationally active asceticism, in mastering the world, seeks to tame what is creatural and wicked through work in a worldly 'vocation' (inner-worldly asceticism). Such asceticism contrasts radically with mysticism, if the latter draws the full conclusion of fleeing from the world (contemplative flight from the world).
>
> The contrast is tempered, however, if active asceticism confines itself to keeping down and to overcoming creatural wickedness in the actor's own nature. For then it enhances the concentration on the firmly established God-willed and active redemptory accomplishments to the point of avoiding any action in the orders of the world (asceticist flight from the world). Thereby active asceticism in external bearing comes close to contemplative flight from the world.
>
> The contrast between asceticism and mysticism is also tempered if the contemplative mystic does not draw the conclusion that he should flee from the world, but, like the inner-worldly asceticist, remain in the orders of the world (inner-worldly mysticism).

In both cases the contrast can actually disappear in practice and some combination of both forms of the quest for salvation may occur. But the contrast may continue to exist even under the veil of external similarity. For the true mystic the principle continues to hold: the creature must be silent so that God may speak. He 'is' in the world and externally 'accommodates' to its orders, but only in order to gain a certainty of his state of grace in opposition to the world by resisting the temptation to take the ways of the world seriously. As we can see with Lao-tse, the typical attitude of the mystic is one of a specifically broken humility, a minimization of action, a sort of religious incognito existence in the world. He proves himself *against* the world, against his action in the world. Inner-worldly asceticism, on the contrary, proves itself *through* action. To the inner-worldly asceticist the conduct of the mystic is an indolent enjoyment of self; to the mystic the conduct of the (inner-worldly active) asceticist is an entanglement in the godless ways of the world combined with complacent self-righteousness. With that 'blissful bigotry,' usually ascribed to the typical Puritan, inner-worldly asceticism executes the positive and divine resolutions whose ultimate meaning remains concealed. Asceticism executes these resolutions as given in the God-ordained rational Orders of the creatural. To the mystic, on the contrary, what matters for his salvation is only the grasping of the ultimate and completely irrational meaning through mystic experience. The forms in which both ways of conduct flee from the world can be distinguished by similar confrontations.

Weber, Max (1967): "Religious Rejections of the World and Their Directions". In: Gerth, Hans H. and C. Wright Mills (eds.): *From Max Weber: Essays in Sociology.* New York: Oxford University Press, 323–359, p. 325f.

The path to salvation provided also a new conception of individual personhood and subjectivity. Working to realize God's will in this world or ascetic retreat from this world and rejection of the pleasures of life allowed to detach from this world and to acquire "real" and "true" life. Crossing the boundary to transcendence, living a life in the name of God, being an empty vessel ready to accept God's will or being God's tool on earth were praised as the path to higher personhood.

Certainly, the transmundane order has, in all human societies, been perceived as somewhat different, usually higher and stronger, than the mundane one. But in the pre-axial-age 'pagan' civilizations this higher world has been symbolically structured according to principles very similar to those of the mundane or lower one. Relatively similar symbolic terms were used for the definition of God(s) and man; of the mundane and transmundane orders—even if there always was a continuous stress on the difference between them. In most such societies the transmundane world was usually equated with a concrete setting, 'the other world', which was the abode of the dead, the world of spirits, and not entirely unlike the mundane world in detail.

These pagan societies, of course, always recognized the moral frailty of man; the failure of people to live up to the prevalent social and moral ideals. However, a conception of an autonomous, distinct moral order which is qualitatively different from both this world and 'the other world' developed only to a minimal degree.

Such homologous conceptions of the transmundane and mundane worlds were very often closely connected with some mythical and cyclical conception of time in which the differences between the major time dimensions—past, present and future—are only mildly articulated.

By contrast, in the axial-age civilizations, the perception of a sharp disjunction between the mundane and transmundane worlds developed. There was a concomitant stress on the existence of a higher transcendental moral or metaphysical order which is beyond any given this- or other-worldly reality.

The development of these conceptions created a problem in the rational, abstract articulation of the givens of human and social existence and of the cosmic order. The root of the problem lies in the fact that the development of such conceptions necessarily poses the question of the ways in which the chasm between the transcendental and the mundane orders can be bridged. This gives rise to the problem of salvation—to use Weber's terminology. The roots of the quest for salvation are given in the consciousness of death and the arbitrariness of human actions and social arrangements. The search for some type of immortality and a way to overcome such arbitrariness are universal to all human societies. In societies in which the mundane and transmundane worlds are defined in relatively homologous terms this search for immortality is on the whole envisaged in terms of some physical continuity. It is usually seen as conditional to the fulfilment of one's concrete obligation to one's group.

Eisenstadt, Shmuel N. (1982): "The Axial Age. The Emergence of Transcendental Visions and the Rise of Clerics". In: *European Journal of Sociology / Archives Européennes de Sociologie 23(2)*, pp. 296ff.

The bridge between the other-worldly and the this-worldly realm consisted in reading God's will or the eternal principles of the cosmos as presented by the holy scriptures revealed to a prophet or an enlightened person. Buddha, Moses, Saint Paul, and Zoroaster, the Jewish prophets or the Greek philosophers and, in secondary axial ages later on, Muhammad, Luther, and Calvin were such charismatic persons who charted a transcendental vision, a principled and eternal system to be contrasted to the pursuit of profane interests and earthly ambitions. Their message of asceticism and conversion attracted followers and engendered communities of faithful who devoted their life to the cause of salvation. Virtuosi like Christian monks or Jewish Rabbis, Muslim Ulema or Buddhist Sangha, Hindu Brahmins or Greek philosophers practiced particular forms of asceticism and studied extensively the Holy Scriptures.

All axial age religions were centered on these Holy Scriptures and the gatekeepers to these scriptures were intellectuals. In most cases the teaching of the transcendental visions resulted in special schools or academies that were not controlled by the political authorities, but by the religious community. Political community and religious community were drifting apart, the prophet and the ruler, the scripture and the sword were parting and even opposing each other as the plough and the sword were parting in conquest societies. The tension between the new intellectual elites who carried and preached a new transcendental vision and the political elites at the center resulted often in a direct challenge to the legitimacy of the ruler. As the this-worldly realm was seen as utterly profane, also political authority which is a part of it was deprived of its sacred nature and seen as a profane office that had to justify its claim with respect to higher ranking eternal principles. Instead of being sacred the ruler is held accountable for his actions in axial age empires.

The profanization of the visible world also opened up the path to a new perspective on nature. The world ceased to be animated by spirits with whom humans could magically interact. Nature was disenchanted and turned into a field of study and measurement, geometry and mathematics were applied to the movements of the cosmos, science as the study of nature began to be separated from the study of the Holy Scriptures.

> The science of Pure Mathematics, in its modern developments, may claim to be the most original creation of the human spirit. Another claimant for this position is music. But we will put aside all rivals, and consider the ground on which such a claim can be made for mathematics. The originality of mathematics consists in the fact that in mathematical science connections between things are exhibited which, apart from the agency of human reason, are extremely unobvious. Thus the ideas, now in the minds of contemporary mathematicians, lie very remote from any notions which can be immediately

> derived by perception through the senses; unless indeed it be perception stimulated and guided by antecedent mathematical knowledge. This is the thesis which I proceed to exemplify.
>
> Suppose we project our imagination backwards through many thousands of years, and endeavour to realise the simple-mindedness of even the greatest intellects in those early societies. Abstract ideas which to us are immediately obvious must have been, for them, matters only of the most dim apprehension. For example take the question of number. We think of the number 'five' as applying to appropriate groups of any entities whatsoever—to five fishes, five children, five apples, five days. Thus in considering the relations of the number 'five' to the number 'three,' we are thinking of two groups of things, one with five members and the other with three members. But we are entirely abstracting from any consideration of any particular entities, or even of any particular sorts of entities, which go to make up the membership of either of the two groups. We are merely thinking of those relationships between those two groups which are entirely independent of the individual essences of any of the members of either group. This is a very remarkable feat of abstraction; and it must have taken ages for the human race to rise to it. During a long period, groups of fishes will have been compared to each other in respect to their multiplicity, and groups of days to each other. But the first man who noticed the analogy between a group of seven fishes and a group of seven days made a notable advance in the history of thought. He was the first man who entertained a concept belonging to the science of pure mathematics.
>
> Whitehead, Alfred North (1967): *Science and the Modern World.* New York: The Free Press, p. 19f.

But even religious communities were not exempted from the dynamics of desacralization. The original ascetic message of the prophets was, after some time, often compromised and watered down, the priests could not resist the pleasures and temptations of this world and gave up their ascetic convictions. The exigencies of organizing and institutionalizing the religious community demanded politics and compromise—at the end the intellectual counter-elite was included in the ruling elite and the tension between the transcendence and profane power collapsed. New movements responded to this decline and attempted to return to the authentic and original message, to retreat from the center and to reject the decadence of the center. The radical zeal of these reformatory movements challenged the center. It responded to this challenge mostly in two ways: on the one hand it showed some flexibility by supporting certain reforms and providing special institutional arenas for the radical virtuosi. A famous example of this is the monastic orders in the European middle ages. The reformatory zeal of religious movements was encapsulated: within the walls of the monastery radical discourse could flourish but it was not allowed to preach this to the commoners. On the other hand the center proclaimed an orthodox way of cultural practice and persecuted all those who publicly refused to accept the orthodox dogma. But even persecution could hardly overcome the faith of those who believed that this world is evil and that heavenly pleasures will wait for the martyrs. The conflict between orthodoxy and heterodoxy reconstructed the axial age tension and it never was completely solved. Even if the center surrendered to the heterodoxy and the previous heterodoxy was turned into a new orthodoxy, the dynamics of trivialization and decadence cannot be stopped in axial age cultures: after some time a new heterodoxy could challenge the center, etc. the tension between orthodoxy and heterodoxy is typical for church like religious organizations that claim to include all mankind regardless whether faithful or heretic. In contrast to churches sects do not claim to represent all mankind. Instead, sects are communities of faithful devotees and true believers mostly without hierarchical structure and open to split and separation if dissent prevails.

> At the outset the actual differences are quite clear. The Church is that type of organization which is overwhelmingly conservative, which to a certain extent accepts the secular order, and dominates the masses; in principle, therefore, it is universal, i.e. it desires to cover the whole life of humanity. The sects, on the other hand, are comparatively small groups; they aspire after personal inward perfection, and they aim at a direct personal fellowship between the members of each group. From the very beginning, therefore, they are forced to organize themselves in small groups, and to renounce the idea of dominating the world. Their attitude towards the world, the State, and Society may be indifferent, tolerant, or hostile, since they have no desire to control and incorporate these forms of social life; on the contrary, they tend to avoid them; their aim is usually either to tolerate their presence alongside of their own body, or even to replace these social institutions by their own society.
>
> Further, both types are in close connection with the actual situation and with the development of Society. The fully developed Church, however, utilizes the State and the ruling classes, and weaves these elements into her own life; she then becomes an integral part of the existing social order; from this standpoint, then, the Church both stabilizes and determines the social order; in so doing, however, she becomes dependent upon the upper classes,

and upon their development. The sects, on the other hand, are connected with the lower classes, or at least with those elements in Society which are opposed to the State and to Society; they work upwards from below, and not downwards from above.

Finally, too, both types vary a good deal in their attitude towards the supernatural and transcendent element in Christianity, and also in their view of its system of asceticism. The Church relates the whole of the secular order as a means and a preparation to the supernatural aim of life, and it incorporates genuine asceticism into its structure as one element in this preparation, all under the very definite direction of the Church. The sects refer their members directly to the supernatural aim of life, and in them the individualistic, directly religious character of asceticism, as a means of union with God, is developed more strongly and fully; the attitude of opposition to the world and its powers, to which the secularized Church now also belongs, tends to develop a theoretical and general asceticism. It must, however, be admitted that asceticism in the Church, and in ecclesiastical monasticism, has a different meaning from that of the renunciation of or hostility to the world which characterizes the asceticism of the sects.

The asceticism of the Church is a method of acquiring virtue, and a special high watermark of religious achievement, connected chiefly with the repression of the senses, or expressing itself in special achievements of a peculiar character; otherwise, however, it presupposes the life of the world as the general background, and the contrast of an average morality which is on relatively good terms with the world. Along these lines, therefore, ecclesiastical asceticism is connected with the asceticism of the redemption cults of late antiquity, and with the detachment required for the contemplative life; in any case, it is connected with a moral dualism.

The asceticism of the sects, on the other hand, is merely the simple principle of detachment from the world, and is expressed in the refusal to use the law, to swear in a court of justice, to own property, to exercise dominion over others, or to take part in war. The sects take the Sermon on the Mount as their ideal; they lay stress on the simple but radical opposition of the Kingdom of God to all secular interests and institutions. They practise renunciation only as a means of charity, as the basis of a thorough-going communism of love, and, since their rules are equally binding upon all, they do not encourage extravagant and heroic deeds, nor the vicarious heroism of some to make up for the worldliness and average morality of others. The ascetic ideal of the sects consists simply in opposition to the world and to its social institutions, but it is not opposition to the sense-life, nor to the average life of humanity. It is therefore only related with the asceticism of monasticism in so far as the latter also creates special conditions, within which it is possible to lead a life according to the Sermon on the Mount, and in harmony with the ideal of the communism of love.

Troeltsch, Ernst (1931): *The Social Teaching of the Christian Churches. Vol. 1.* London: George Allen & Unwin; New York: The Macmillan Company, p. 331f.

Circulation: Markets, Capital Cities, and Vertical Mobility

The transportation system of empires and the pacification of its territories created large spaces for the circulation of goods. Still many economic transactions were performed without money, but long-distance trade was much increased, coins and currency standardized money exchange for the entire empire and made prices comparable, markets extended beyond the confines of cities and villages and set prices with little regional variations, shortages in one region could be compensated by long-distance transportation, complex economic operations like money lending and loans started. Money was gradually taking over the role of the key axis of social structure, a role that ownership of land and command over people had before. But long-distance trade was not confined to the inner space of the empire. It extended also to the outside and, sometimes, reached out to other empires. Arab dhows sailed to the East African coast and India, Chinese junks arrived at Indian ports etc. The Roman Empire and the Chinese Han Empire were furthermore connected by a regular caravan route by which mainly precious metals from the west were exchanged against silk from China.

The growth of circulation and mobility was, however, not limited to the level of goods. The mobility of people, too, increased considerably. Officials from the capital were appointed to positions in the provinces and called back to report or to be promoted. Armies were moved from one frontier of the empire to another, merchants traveled with their wares over long distances, slaves were traded from newly conquered territories to the center, convicted citizens from Rome were banned to some remote provinces, people from the provinces moved to the capital city in search of fortune, etc. But the city did not attract people only because they expected opportunities. It was also the excitement provided by entertainments like circus and theater that fascinated those who left the dullness of the provinces where everything was regular and expectable. In empires the novelties and excitements occur at the center. Even in contemporary societies this fascination of the big city in particular for young people from the provinces is very much alive.

The dynamics of circulation in empires exhibits a clear centripetal direction. Many people wanted to stay in the capital city and whoever moved to the provinces did this for returning later back to the capital city.

The capitals attraction of persons and goods was counteracted by the attempt of the center to control the provinces and to expand the frontier. In order to achieve this control empires dispose of much more free-floating resources (people, goods, money) than conquest societies. Since the political authority of the center was much less based on violence, terror, and pillage than was common in conquest societies, its legitimacy was much greater. Roman armies marching through the empire did not live on pillaging the villages they passed by, but on food provisions they carried with them. The high costs of the standing army, however, strained the budget of the ruler. Hence the taxation system was of paramount importance for the budget of the ruler and frequently the invention of new taxes was to cover the ever growing expenses.

More striking than the increase of horizontal, that is, territorial mobility is the vertical mobility and the opening of rank society. Certainly there were still the landed aristocratic families who were the descendants of the old warrior classes and who did, of course, not work for a living but enjoyed a luxurious life and were influential on the political level. There were still the urban middle classes of craftsmen and small merchants, etc. who were sometimes organized in guilds and corporations that had a privilege to trade a defined good or to perform a certain craft. The corporation or guild regulated the access to their membership, set just prices, and thus prevented competition.

But in spite of these monopolies for crafts and trade considerable mobility and amazing careers were possible. This extended up to the position of the emperor himself. Among the Roman emperors were people from the provinces, successful generals, former criminals, superrich landowners, etc. The position of the emperor was not strictly regulated by law, and the core problem of succession was never clearly solved as in a traditional monarchy. The son of the emperor was not in every case the obvious successor. Instead, complex influences such as the last will of the previous emperor, military might, support by the old families, the opinion of the Senate and intrigues at the imperial court were involved in deciding about the person of the new emperor who claimed to be "princeps among equals."

Representation: Literature, Theater, and Patronage

The upper classes in empires excelled not only by an extraordinary luxurious life style but also by literacy. While, in conquest societies, reading and writing was a professional matter of scribes, in empires almost every member of the upper classes was able to read and write. Some empires like the Inca Empire, however, did not dispose of a complex writing. This resulted in a vast readership for special writings that did not serve practical purposes but aimed at entertainment and general reflection. Narrating stories was shifted from the oral practice to the written representation. Literature in the proper sense was born. The works of literature were set free from practical functions and became the object of a special discourse on aesthetics. Classical writings were canonized as an ideal to strive for.

The relationship of literature to reality differs from the relationship of written reports or commercial contracts to reality. Reports can be proved wrong by reality, contracts can be broken by real actions, but there is neither such thing as wrong literature nor does literature claim certain actions to be performed or others to be refrained from. The mode of literary representation is, instead, *fictional*, that is, it imagines a possible reality and its readers know about this fictionality. In literature the relationship between textual representation and referential reality is not a natural and immediate one. The representation does not presuppose that there is a real reference. But when reading a literary text, we do neither have to assume that it is just fantasy. We indulge in the fictional reality presented by the text.

A similar awareness of the difference between representation and reality holds true for theatrical performances. In distinction to rituals, theatrical performances do stage a reality and the audience knows that the people on stage are actors and the performance is not real in the everyday sense of the term. A murder on the theatrical stage will not be persecuted afterward. But at the same time the audience wants the performance to be a believable illusion, they want to be captivated by the staged reality, they want to forget that it is just a blunder and they show real emotions if the performance is able to produce this fusion of theater and reality. Literary fiction and theater project a reality but there is no way of checking the reality to which they refer. Both modes of representation differ also from the constitutive power of ritual performances. Rituals create a reality that cannot be questioned or debunked as illusionary. After marrying a person we cannot declare this marriage to have been just a theatrical performance and thus being invalid.

In distinction to literary fiction and theater written texts can also be critically examined with respect to the reality to which they refer. Instead of the longing for illusion the reader is here moved by the suspicion of treacherous and illusionary assumptions made by the text. Because a written text is disembodied and cut off from its original context, its meaning is more open to interpretation than oral speech. But this interpretation of a text is also much easier if the text exists in a written form in front of the reader. We can read a text again and again, we can stop and think about its hidden meaning, we can look for clues and hints, and we are not bound to follow the sequential turn-taking of oral communication. No other person is responding, the reader is in a dialogue with an invisible and unknown other, and this dialogue with the text is patterned by rules that differ from the turns between speech and response. The temporal sequence is replaced by the timeless coexistence of signs in a text and the text, devoid of time and location, becomes the object of thinking. This situation fosters *second order reflections* that search for timeless, abstract, and translocal principles hidden in the texts. Writing about these principles imagines a world that is structured by an invisible eternal order behind the surface of everyday life. Thus the interpretation of written texts by intellectuals can give rise to the axial age divide between the profane and the transcendental realms that we have discussed above.

As widespread as literacy was among the upper classes of empires, the written text of the time were rare and precious. Their reproduction was slow and expensive. Many young scholars spent most of their time in copying scriptures by hand. The importance of copying is also mirrored by a special intellectual ethic that praised the founding fathers and treated their works as ideals to be followed. Nobody could surpass and exceed the classics. The best literature and art could do was to repeat the classical form and to return to the classical origins. Thus representation in traditional art and literature did not only refer to a (fictional) reality but also to a classical model or exemplar. This classicism climaxed sometimes in cultural movements of renaissance: under the rule of the emperor Hadrian Roman art returned programmatically to its Greek origins, the Sung had their Confucian revival, etc.

But empires did not only develop new modes of representation by written texts, by theatrical performances, and by pictorial art. Its large population required also new modes of representation in the fields of politics.

The authority of the ruler in ancient empires differs from the authority of conquerors. It did not just depend on brute force and violence, on charismatic presence or on descent. Some of the rulers were deified (e.g., the Inca or some Roman emperors) or combined religious and political leadership in their person (e.g., the Caliphate). The majority of rulers in ancient empires however needed the blessings or the benevolence of the deities in order to keep their authority even if they were called the "son of heaven," etc. If the signs of the heaven were gloomy the authority of the Chinese as well as of the Roman emperor could be seriously questioned.

Even more important than the blessings of heaven was, however, the support by the majority of the people. As insane and idiosyncratic as some of the Roman emperors may appear to us their authority was legally based on the consent of "Senatus populusque Romanus" (Senate and people of Rome). To rule an empire the ruler needed a council that represented his people and in particular the elites. Here a new mode of representation enters the arena. Representing a group of people does not refer to objects or sacred forces as symbols do. Instead, *political representation* refers to the will of people who are, for contingent reasons, unable to articulate their will and voice in the political arena. The most elementary form of political representation is the relationship between patrons and clients, the most complex is the party system of modern representative democracies. Clients are unable to put forward their will in political arenas and sometimes they even do not know precisely what their actual interests are. But they trust their patron who is able to participate in political debates and to voice the interests of his clients when it matters. Trust, not authority, power or charisma is the basis of this political representation by patronage. Sometimes the clients pay the patron for his service, in most cases, however, the patron's gains are political ones. By a large clientele he can increase his own political power. Usually clients chose their patron on their own decision, but sometimes the individual person is bound by the collective decision of her group for a patron. In more elaborated systems of patronage the patron is elected to represent the group in a particular institutional arena. Patronage gets increasing salience as the size of the total population grows and the presence of all citizens in the council of the ruler becomes impossible. Thus patronage and clientelism becomes an integrative institution crossing the boundaries of kinship and locality, but organizing loyalty and politics below the level of centralized authority.

References

Eisenstadt, Shmuel N. (1986, ed.): *The Origins and Diversity of Axial Age Civilizations.* Albany: State University of New York Press.

Eisenstadt, Shmuel N. (1982): "The Axial Age. The Emergence of Transcendental Visions and the Rise of Clerics". In: *European Journal of Sociology / Archives Européennes de Sociologie* 23(2), 294–314.

Jaspers, Karl (1965): *The Origin and Goal of History.* New Haven; London: Yale University Press.

Mann, Michael (1986): *The Sources of Social Power. Vol. I.* Melbourne: Press Syndicate of the University of Cambridge.

Troeltsch, Ernst (1931): *The Social Teaching of the Christian Churches. Vol. 1.* London: George Allen & Unwin; New York: The Macmillan Company.

Weber, Max (1967): "Religious Rejections of the World and Their Directions". In: Gerth, Hans H. and C. Wright Mills (eds.): *From Max Weber: Essays in Sociology.* New York: Oxford University Press, 323–359.

Whitehead, Alfred North (1967): *Science and the Modern World.* New York: The Free Press.

Territorial States

Bernhard Giesen

Production: The Turn Inward and the Beginnings of Modernization

Imperial expansion reaches its limits when, beyond its frontiers, it encounters no "barbarians" but, instead, other states which claim civilizational parity and try to expand their rule by military means. Faced with another power that mirrors the empire the risks of conquest increase for both sides. In response to these rising costs of imperial expansion for both sides, the diffuse frontier zone was gradually turned into a clearly marked boundary that is mutually agreed upon and protected by legal contracts and criss-crossing alliances. This transformation occurred in early modern Europe.

Turning frontier zones into defined boundaries does not prevent wars. There were, indeed, a lot of wars in early modern Europe and many of them were propelled by the motives of military conquest or religious zeal. But after the Thirty Years War that had devastated and depopulated Central Europe in the seventeenth century these wars had increasingly to be prepared by special diplomatic procedures that were performed by specially appointed officials who enjoyed immunity in the territory of the enemy. Even more important, wars had to be justified by particular reasons such as broken contracts, provocative moves of the enemy, liberation from insupportable tyranny or disputes about succession and heritage. Since the early modern age war ceased to be the normal situation at the boundaries of territorial states in Europe.

B. Giesen et al., *Human Societies*, Cultural Sociology,
https://doi.org/10.1007/978-3-032-07697-7_7

This transformation of frontier zones into legally defined and commonly agreed boundaries profoundly affected the economy. The traditional way of generating additional power by conquering new territories and including the defeated people into the population of the empire did no longer yield significant gains. People beyond the boundary could not be treated as inferior pagan savages in need of civilization. Instead, Europe consisted of a range of states that faced each other on an approximately equal footing and whose princely rulers were frequently related to each other by family ties.

In this situation the imperial politics of conquest could, supported by fast maritime transportation, turn to other continents where "wilderness" and "pagan savages" seemed to require missionaries and conquerors. Sailors and soldiers from Spain and the Netherlands, from Britain and France ventured out to search for fortune and treasures. They prepared the path for merchants and administrators who marked territorial claims and established outposts. Shipbuilding was promoted and large overseas ports were constructed. Fast sailing ships and guns, ruthlessness and the lure of riches turned, within two centuries, the world outside Europe into colonies of European powers.

In distinction to the expansion of many ancient empires, this politics of colonialism did, however, hardly result in the complete civil inclusion of the colonized people. Millions of the indigenous population in the Americas were killed by plagues and unbearable working conditions in plantations and mines. They were replaced by black African people who were enslaved and shipped across the Atlantic. The early modern colonial system continued the atrocities of conquest but shifted it to overseas territories that were shielded from the eyes of its citizens. Even later on, after the advent of modernity, the social barriers between the conquerors and the colonized people remained stronger than in most ancient empires: official intermarriage was low, chances for careers were limited for the colonized even if they had accepted religion and education of the colonizers. Colonialism allowed the imperial center to exploit the periphery overseas and, at the same time, to be safe from the risk of barbarian invasions that had imperiled the heartlands of the ancient empires, for example, Rome and China.

But the turn toward colonialism was not the only way of coping with the shrinking chances of conquest within Europe. Faced with demarcated and fixed boundaries, the princely ruler could turn inward to discover resources that had been ignored before. Forests could be cut and swamps could be dried in order to gain additional arable land, canals and roads could be constructed to ease transportation, migrants from abroad could be invited to settle as farmers in the newly gained land. The politics of religious homogeneity established after the confessional wars between Protestants and Catholics in the seventeenth century forced many highly skilled citizens to leave their country of birth and to take refuge in those states that allowed them to practice their religion. This religiously motivated migration provided new opportunities to increase the working population and thus to boost the economic productivity.

But immigration was not the only strategy of increasing the work force without conquest. Equally important and effective were repressive policies against unproductive parts of the population such as vagabonds, beggars, the jobless, and masterless people who were despised, yet tolerated in ancient societies. The new politics of modernization wanted the population to be sedentary and accountable, productive and taxable. Beggars and hobos, gypsies and thieves had no place in the emerging society of working citizens. This politics was fostered by the religious convictions of reformatory Protestantism. It commanded an ascetic attitude toward this worldly pleasure and claimed that inner-worldly, that is, economic success of a person hinted at his or her choseness for salvation in heaven. Thus work was converted from burden and nuisance into vocation and virtue. Supported by the Protestant ethics the middle classes in early modern territorial states devoted their life to working. Those who refused to follow this path were removed from the public spaces, expelled from the territory or imprisoned in working houses. Thus the early modern territorial state could increase its work force without necessarily increasing its population.

The growth of the workforce, however, was not the only factor that yielded better resources for taxation. It was also the higher productivity of work based on technological inventions and improvements that were fostered, supported, and promoted by the politics of modernization. New technologies such as printing, cannons, or weaving machines—as different as they are—were breakthroughs in efficiency, energy expenditure, and standardization. After some time the princely states did compete not only with respect to military might and courtly splendor but also on the field of technologies. Modern technologies provided relative advantages in the production of goods that could be sold abroad at high prices. Thus the princely ruler as well as merchants and manufacturers were joined in a common interest in modernization.

Classification: Monarchs, Aristocrats, and the Bourgeoisie

From the seventeenth century onward kings and the bourgeoisie were united in their quest for modernization. The king needed officials for his administration and merchants for taxing their wealth. The bourgeoisie, in turn, needed legal protection by the state and new fields for making profits. The bourgeoisie was granted privileges of trade and production, the colonial empires shipped the products of their colonies into the centers and, in return, sold wares to the colonies. The wealth of the merchants increased rapidly and exceeded the ones of most aristocrats. Cities like Venice, Florence, Amsterdam, Bruges, Nuremberg, and Augsburg were not only flourishing centers of commercial life but gained also a certain political autonomy with respect to kings and emperors. Their patrician leaders presented themselves in aristocratic splendor, supported the arts and hired armies to wage war against each other or the ruler. Occasionally ultrawealthy merchants like the Fugger in Augsburg (Germany) lent money even to kings and emperors to fund their colonial expeditions.

The rise of relatively autonomous cities and their merchant class was, however, only the most spectacular aspect of the bourgeoisie's way to power and domination. Equally important was the rulers attempt to widen and tighten the control over land and people by trained and appointed officials. Here it was not money and markets, but education and administration that supported the rise of the bourgeoisie. The administration of the territorial state was increasingly handed from noblemen who got their position by descent to bureaucrats who got their position by training and examination. Lineage and land was replaced by education and examination as the prime criteria for assigning social status.

Holding an office was no longer limited to the top officials of the princely administration. Also minor tasks were appointed only to those who had passed an examination after receiving a special education. Thus the number of positions in the administration grew. A new class of educated officeholders and professionals emerged. This new class separated itself from the traditional local middle classes. They tended to despise the narrow-mindedness of craftsmen and shopkeepers. While these insisted on local roots and traditional privileges the new group of administrators was devoted to the princely ruler and to the impersonal rule of the law. In order to enhance the progress of science and technology many European rulers established academies and supported learned associations. Discourse and proceedings of these scholarly institutions were based on scientific objectivity and impersonal reasoning. Here, too, descent and tradition were devaluated as a condition of membership.

The traditional aristocracy whose power was based on land and sword found itself gradually in an odd situation. In the old feudal system, their military might and economic resources had granted them a position that rivaled the power of the king whose authority was occasionally diminished to sheer decorative function. At the height of feudalism in the eleventh century Europe was scattered into many territories owned, exploited, and ruled by largely independent feudal lords. Although, even in this period of decentralized feudal rule, the king claimed an authority that was not only based on his private land and power, the political system was far from being state-like in the modern sense of the term.

The rise of the European monarchy (i.e., the exclusive rule of a single prince in the kingdom) had to cope with the traditional power of the landed aristocracy who, for obvious reasons, resisted the monarchic claim to superior authority. As long as the nobility resided on their lands and were able to wage war with the central authority by their own troops, this resistance had a chance to succeed. Such a resistance led to several large conspiracies against the monarch, the most famous of them being the Fronde in seventeenth-century France. The French, English, and Spanish monarchs prevailed, however, against these uprisings of the aristocracy by relying on the military might of a large standing army of professional soldiers. In China the emperors of the Tang dynasty had disempowered the landed aristocracy already in the ninth century and replaced them by trained and examined officials, the Mandarins. In Japan, by contrast, the aristocracy was able to take over the center of political authority and to reduce the Tenno, that is, the emperor to a merely symbolic figure. The central authority was shifted to the Shoguns who advocated the cause of the nobility.

But the decline of the aristocratic power in most European states (Poland was an exception) was not only due to military defeat. It was also fostered by the nobility itself who responded to the attraction of the royal court and the distraction of urban life by moving from their rural homelands to the royal court and to the capital city. The princely court with its artistic splendor, theater and music, festive parties and extensive social life included the aristocracy and turned them from warriors into courtsmen deprived of their original resources of power.

This change profoundly affected their canon of virtues and their idea of honor. Valor and heroic vocation of the knight was replaced by elegance and courtesy, manners and wit, refined taste and relaxed generosity. Conversation became an art of its own, dancing, promenading, and dining were to display ease, graciousness, and elegance.

> [...] I judge the principall and true profession of a Courtyer ought to be in feates of armes, the which above all I will have hym to practise lively, and to bee knowen among other for his hardinesse, for his acheving of enterprises, and for his fidelitie toward him whom he serveth. And he shall purchase himselfe a name with these good condicions, in doing the dedes in everie time and place: for it is not for him to feint at any time in this behalfe without a wonderous reproche. And even as in women honestye once stained dothe never retourne againe to the former astate: so the fame of a gentleman that carieth weapon, yf it once take a foile in any little point through dastardlines or any other reproche, doeth evermore continue shameful in the worlde and full of ignoraunce. Therefore the more excellent our Courtyer shalbe in this arte, the more shall he bee worthy praise: albeit I judge not necessarye in hym so perfect a knowledge of thynges and other qualities that is requisite in a capitaine. But because this is overlarge a scope of matters, wee wyll holde oure selves contented (as wee have sayde) with the uprightnesse of a well meaning minde, and with an invincible courage, and that he alwaies shew himself such a one: for many times men of courage are sooner knowen in small matters then in greate. Often times in daungers that stande them upon, and where many sometime eyes be, ye shall see some that for all their hearte is dead in their bodie, yet pricked with shame or with the company, go forwarde (as it were) blindfield and do their dutie. And God knoweth bothe in matters that little touche them, and also where they suppose that without missynge they may convey themselves from daunger, how they are willing ynough to slepe in a whole skinne. But suche as think themselves neither marked, seen, nor knowen, and yet declare a stout courage, and suffer not the leaste thyng in the worlde to passe that maie burthen them, they have the courage of spirite whiche we seke to have in our Courtyer. Yet will we not have him for al that so lustie to make braverie in woordes, and to bragge that he hathe wedded his harneys for his wife, and to threaten with suche grim lookes, as we have seene Berto do oftentimes. For unto suche maie well be saide that a worthie Gentlewoman in a noble assembly spake pleasauntly unto one, that shall be namelesse for this tyme, whome she to shewe hym a good countenance, desired to daunce with her, and he refusing both that, and to heare musick and many other entertainmentes offred him, alwaies affirming suche trifles not to be his profession, at

> last the Gentlewoman demaundyng him, What is then your profession? He aunswered with a frowning looke: To fight.
>
> Then saide the Gentlewoman: Seing you are not nowe at the warre nor in place to fight, I woulde thinke it beste for you to bee well besmered and set up in an armorie with other implementes of warre till time wer that you should be occupied, least you waxe more rustier then you are.
>
> Thus with muche laughinge of the standers by she left him with a mocke in his foolishe presumpcion. He therefore that we seeke for, where the enemies are, shall shewe himselfe moste fierce, bitter, and evermore with the firste. In everie place beside, lowly, sober, and circumspecte, fleeing above all thinge bragginge and unshamefull praising himself, for therewith a man alwaies purchaseth himself the hatred and yll will of the hearers.
>
> Castiglione, Baldassare (1900): *The Book of The Courtier. From the Italian of Count Castiglione: Done into English by Sir Thomas Hoby. Anno 1561. With an Introduction by Walter Raleigh.* London: David Nutt in the Strand, pp. 48–50

Faced with the loss of military power the nobility had to keep the rising middle classes at a distance. It constructed this distance by focusing on descent: you have it by birth or you do not have it. There is no way of learning, imitating, or copying the ease and the manners that noblemen got by birth. Whoever calculates his advantage, whoever tries to find a cheap way of handling marks himself as an ordinary person, as a "pedant" in distinction to the generosity or graciousness displayed by the French "gentilhomme" or to the British "gentleman." Thus the decline of military power was associated with constructing boundaries around the aristocracy as a community of manners.

> In the process of transformation and innovation that we designate by the term "Renaissance," what was regarded as "fitting" and "unfitting" in human intercourse no doubt changed to a certain degree. But the rupture is not marked by a sudden demand for new modes of behavior opposed to the old. The tradition of *courtoisie* is continued in many respects by the society which adopts the concept of *civilitas*, as in *Civilitas morum puerilium*, to designate social "good behavior."
>
> The increased tendency of people to observe themselves and others is one sign of how the whole question of behavior is now taking on a different character: people mold themselves and others more deliberately than in the Middle Ages.
>
> Then they were told, do this and not that; but by and large a great deal was let pass. For centuries roughly the same rules, elementary by our

standards, were repeated, obviously without producing firmly established habits. This now changes. The coercion exerted by people on one another increases, the demand for "good behavior" is raised more emphatically. All problems concerned with behavior take on new importance. The fact that Erasmus brought together in a prose work rules of conduct that had previously been uttered chiefly in mnemonic verses or scattered in treatises on other subjects, and for the first time devoted a separate book to the whole question of behavior in society, not only at table, is a clear sign of the growing importance of the question, as is the book's success. And the emergence of related writings, like the *Courtier* of Castiglione or the *Galateo* of Della Casa, to name only the most well-known, points in the same direction. The underlying social processes have already been indicated and will be discussed in more detail later: the old social ties are, if not broken, extensively loosened and are in a process of transformation. Individuals of different social origins are thrown together. The social circulation of ascending and descending groups and individuals speeds up.

Then, slowly, in the course of the sixteenth century, earlier here and later there and almost everywhere with numerous reverses until well into the seventeenth century, a more rigid social hierarchy begins to establish itself once more, and from elements of diverse social origins a new upper class, a new aristocracy forms. For this very reason the question of uniform good behavior becomes increasingly acute, particularly as the changed structure of the new upper class exposes each individual member to an unprecedented extent to the pressure of others and of social control. It is in this context that the writings on manners of Erasmus, Castiglione, Della Casa, and others are produced. People, forced to live with one another in a new way, become more sensitive to the Impulses of others. Not abruptly but very gradually the code of behavior becomes stricter and the degree of consideration expected of others becomes greater. The sense of what to do and what not to do in order not to offend or shock others becomes subtler, and in conjunction with the new power relationships the social imperative not to offend others becomes more binding, as compared to the preceding phase.

Elias, Norbert (1978): *The Civilizing Process. The Development of Manners*. New York: Urizen Books, p. 79f.

The middle classes, in turn, strived to be accepted by the prestigious center of social life. Even if its members had made a fortune they were still largely excluded from the aristocratic community of manners and honor. In many cases they tried to imitate courteous manners, to learn it by the book, and even applied to the prince for nobilitation in exchange for merit or money. At the end of the eighteenth century many of them had succeeded. In 1830 most French citizen with noble titles could not claim a noble descent of more than half a century. The traditional aristocracy despised the freshly nobilitated citizens as "gens de robe" (people of office) in distinction to the "noblesse d'épée" (sword carrying nobility).

The center of court life was, of course, the monarch. His (or rarely: her) claim to rule by the grace of God explicitly denied any need to base sovereignty on the peoples vote and consent. It was the ruler who defined his people and not the people who elected the ruler. The relationship between monarch and people was imagined as a parental or patriarchical one. The monarch was the symbolic founder of the state, the embodiment of its unity, the source of law and justice, the graceful provider of education and care. In some cases he was even entitled to spend the first night with every bride in his population thus indicating that he was the symbolic father of his people. While the administration below the prince was increasingly controlled by impersonal laws, the authority of the European monarchs was absolute. Even if the ruler was not an impressive person (and many of them were not), his presence had still a charismatic radiation. It commanded unconditional attention and defined the center of the state. Many people addressed humbly to the monarch for support or favors and the prince granted graces, privileges, and fortunes. He was not obliged to do so but this arbitrariness is the essence of grace and generosity.

> And just as the majority of the nobility were transformed from knights into court *seigneurs* and *grands seigneurs*, the kings too were transformed in the same way. Francis I was still a knightly king, *le roi Chevalier*. He was fond of tournaments and hunting; war for him was a splendid knightly game in which he put his life at stake. For this was required by chivalrous-aristocratic convention, by his honour, and he felt himself bound as king by this law of knightly behaviour like every other knight.
>
> The case was still similar with Henry IV who, when a Huguenot leader and a great vassal of the kings of France, on receiving the news that their opponents were preparing for war, offered to settle the matter by personal combat with the enemy leader, the Duc de Guise: 'Inequality of rank shall not deter me.' One against one, two against two, ten against ten or twenty against twenty, they would fight with the weapons that were customary in an affair of honour between knights. Thus he spoke. After attaining power he embodied the transition between the late-chivalrous type of king and the court-aristocratic type that was to find its first perfect representative in Louis XIV, who no longer, like Henry IV, rode into battle at the head of his nobles. He had his wars fought more and more by generals with paid troops,

> and even if he occasionally exposed himself to fire, he was hardly accustomed to physical warfare. Tournaments too, under Louis XIV, entirely lost their character as personal, man-to-man combat. They became a kind of court game. And if we are looking for an example of how much the king had now become a court aristocrat in his behaviour, but how much, at the same time, his person carried a special weight within court society that distanced him from the others, we might consider this picture of a knightly game that took place in 1662 under Louis XIV [...]
>
> Elias, Norbert (1984): *The Court Society*. New York: Pantheon Books, p. 148f.

Unlike the marcher lords of conquest states, however, who moved with their warriors through their realm, the European monarchs rarely left their huge palaces where court life unfolded and flourished. In France, Spain, Britain, and, later on, in Austria and Russia the royal court included thousands of persons—servants, minstrels, officials, as well as noblemen and visitors, musicians and writers, artists and artisans, and it consumed large parts of the state revenues.

The European courts rivaled each other for splendor and attraction. Their display of pageantry and luxury was, however, not due to the princes' personal hedonistic desires but to the structural tension between the crown and the nobility and the resulting urge to attract the nobility from their homelands into the capital and to bind them by splendor and glamor.

The majority of the population, however, belonged neither to the class of wealthy merchants or the group of educated administrators nor were they included into the nobility. They were peasants, servants, craftsmen, and homeless people. Although slavery was shifted to the colonies, the conditions of their life did not differ very much from the one that common people had to face in ancient empires: hard work and scarce benefits, lingering violence and tough punishment. The peasant serfs could be sold, they were not allowed to move from their allotted farmlands, they could marry only by permission of their master etc.

Circulation: Taxation, Mercantilism, and Printing

The incessant and huge expenditures of the court as well as the costs of a standing army set the finances of the princely states under constant pressure. The prince and his bureaucracy had to cope with debts and deficits. They did so partly by decoupling the currency from its original metal value, that is, by inflation and by introducing new coins and even currencies. The French kings were, hence, called “les rois faux-monnayeurs” (the fake money kings). But the princes could also borrow money from their wealthy citizens via loans or raise ever new taxes on economic transactions, consumption, and property. A range of princely tax collectors assured that the state profited from the economic growth and the expanding market exchange. Thus the basic structure of state finances was set. Even today, the relationship between expenditures, taxes, and loans patterns the economics of the state and the core question is whether the investive expenses of the state (public expenditure) stimulate more economic growth in the long run than it suffocates on a short-term basis.

Most important, however, was a new policy that aimed at closing the state borders for products from abroad by raising high customs. Instead of importing goods from abroad and thus shifting the profits to foreign producers the citizens should be encouraged to buy goods made in the homeland. In order to ensure the necessary supply of goods and services and to foster economic self-reliance, new crafts and industries were encouraged and established. While as few goods as possible should be imported, exports were supported. Thus, for the first time in history economic policies were deliberately set and implanted. The sound principle of spending less money than having income was transferred from the private household to the state. The state was conceived of as an economic actor in a field of international competitors. This politics of *Mercantilism* seemed sound and supported the crafts and manufacturers at home, but it impeded international trade.

One of the most important breakthroughs that occurred in early modern Europe and since then has changed the world of communication is the printing of books. In distinction to writing it belongs more to the sphere of circulation than generating a new mode of symbolic representation. Copying the bible by hand took around ten years. Printing reduces the immense cost of a handwritten book copy to a fraction of it and it made books accessible to a much wider readership. Not only the happy few who were granted the privilege of getting access to the famous libraries of ancient societies but almost every citizen could enjoy the pleasures of reading. Between 1400 AD and 1500 AD the total number of books in Europe jumped from less than 10,000 to an amazing 10 million (estimates). Books as material objects were no longer an immensely valuable semi communal property, instead they could be privately owned. Their texts, by contrast, were no longer precious secrets enjoyed by privileged individuals, but they were publicly accessible. The size of books decreased considerably allowing for easy transportation. The path to the private library was open and publication got its contemporary meaning of general access for anonymous readers.

This also changed the range of the public sphere. Before printing, the public sphere coincided largely with public spaces. It referred to those places and spaces that were accessible for all people present and that did not shield presence and behavior from being exposed to or being noticed by other persons present in the location. Printing decoupled the public sphere from public spaces and bodily presence. It was the text and not the behavior that was noticed publicly. Addressing an anonymous readership by printed books or journals creates, thus, a translocal public sphere in which every reader knows that many other absent persons are reading the same news at roughly the same time.

> The bourgeois public sphere may be conceived above all as the sphere of private people come together as a public; they soon claimed the public sphere regulated from above against the public authorities themselves, to engage them in a debate over the general rules governing relations in the basically privatized but publicly relevant sphere of commodity exchange and social labor. The medium of this political confrontation was peculiar and without historical precedent: people's public use of their reason (*öffentliches Räsonnement*). [...]
>
> To be sure, before the public sphere explicitly assumed political functions in the tension-charged field of state-society relations, the subjectivity originating in the intimate sphere of the conjugal family created, so to speak, its own public. Even before the control over the public sphere by public authority was contested and finally wrested away by the critical reasoning of private persons on political issues, there evolved under its cover a public sphere in apolitical form - the literary precursor of the public sphere operative in the public domain. It provided the training ground for a critical public reflection still preoccupied with itself—a process of self-clarification of private people focusing on the genuine experiences of their novel privateness.
>
> Habermas, Jürgen (1989): *The Structural Transformation of the Public Sphere. An Inquiry into a Category of Bourgeois Society.* Cambridge: The MIT Press, pp. 27, 29

While handwriting had to rely on respect for tradition and careful execution in order to copy the text of the book precisely, printing replaced what, before, was a matter of virtue and craftsmanship by mechanical standardization. The very procedure of printing ensured accurate reproduction.

This devaluation of copying had a strong impact on intellectual and scholarly virtues. Veneration of tradition and repetition of classical texts lost their position as prime references for scholars. Because reproduction was safeguarded by printing, intellectuals could reflect on the text and the traditional text could become the object of critical comment. Criticism was gradually turned from heretical deviance into scholarly virtue. Finally, at the beginnings of modern age intellectuals imagined themselves as critics of tradition, as discoverers of the unknown and creators of the new. Tradition was increasingly seen as obscure and limiting, progress was to be made toward enlightenment and emancipation and the critical discourse in the public sphere was to be the vehicle of this progress. Thus the ever accelerating spiral of public communication that is a hallmark of modernity took on its move. To assume that public communication may improve things is not self-evident: other cultures trust silence and secrecy more.

Printing fundamentally changed also the relationship between the author of a text and its readers. In ancient cultures the author of the handwritten text remained frequently anonymous, whereas the author of the printed text acquired a strong visibility and identity in distinction to the multitude of its readers. Communicating with an anonymous readership became the regular relationship.

This also affected politics and administration. The ruler and his officials could register and control their citizens by standardized printed forms and passports. Without a passport no citizen was able to cross borders, without being registered on printed forms no citizen was perceived by the state as the carrier of rights and duties. Handwritten letters of privilege for individual persons or corporations were replaced by standardized printed forms that assumed that nobody was a singular case but that every citizen was comparable to others. Thus the early modern state preluded the democratic creed of equality.

Representation: Science and Illusionary Art, Nature and Progress

The new modes of representation in early modernity were not so much concerned with representing the sacred, with representing verticality or with representing a demonic outside. There was, of course, a strong representation of vertical rank in early modern society, there was the political representation of the kingdom by the natural person of the king, there was still witch hunting and the adoration of sacred statues and images. But these well-known contents of representation were completed by a new reference. This new reference was nature. Nature was seen as an infinite and continuous realm created by God and testifying his wisdom, reigned by eternal laws and open to discovery and use by human beings. Nature mirrored the mind of God and became itself increasingly an embodiment of the sacred.

But the salience of nature as the prime reference of representation engendered also its obfuscation. Unlike the profane world that, in a premodern perspective, was just out there and easy to account for, nature participated, in a modern perspective, in the invisibility of God. Instead of looking for ways to disclose the will of God as premodern scholars did, the modern scientist searches for discovering the laws of nature. The laws of nature were not obvious and self-evident in the way the phenomena of our everyday world appear to us. Instead, they have to be discovered by methodical procedures and the question whether these discoveries are true or not takes center stage. Modern science is moved by the quest for truth and tries to fit the scientific representation to its natural references by constantly raising this question.

Nature is imagined as an immense mechanical machine that moves predictably according to fixed rules like a clock. We are able to perceive the movements of the clock but its laws are yet to be discovered. The principle of impersonal validity that is at the core of modern legal systems holds true also for modern science: laws of nature claim intersubjective truth. They must be confirmed by empirical observations that, in principle, every reasonable human being can perform. The manifold phenomena of nature were not just described with respect to their variation—all of them testifying the wisdom of God the creator—but they were measured by numbers and—ideally—reduced to mathematical formula. Numbers abstract from the manifold differences between phenomena and focus on one single attribute that can be measured and easily compared to other cases. Weight and length, time spans and quantities are represented by numbers and these numbers can be compared. This quick and accurate comparison of differences is the great attraction of representation by numbers. Instead of looking for similarities between the manifold of phenomena as the premodern scholarship did, modern science focuses on differences and tries to order these differences. The prize of this comparison of differences, however, consists of disregarding the particular manifolds of objects, persons, and events. Quantitative relationships cannot tell a story like texts can do and they cannot imagine faces like paintings can do.

Premodern scholarship dealt with numbers from a largely different perspective. Numbers had a coveted meaning and their relationship disclosed future events or demonic threats—similar to today's astrology that still tries to find out fate and future by exploring the constellations of planets. In modern science, by contrast, numbers are mainly used to measure empirical phenomena of nature. But measurement by numbers was not limited to the empirical description of nature alone. The early modern state, too, needed information about its citizens, their income and property, information about the economy and information about its own fiscal situation. Therefore the emerging "science of the state" used increasingly statistical accounts of the population and fiscal budgets of costs and income. The reign of the number as the prime mode of representation began also for the social realm. This extends, of course, in particular to the realm of economic behavior.

> The extraordinary importance of the highest possible degree of calculability as the basis for efficient capital accounting will be noted time and again throughout the discussion of the sociological conditions of economic activity. It is far from the case that only economic factors are important to it. On the contrary, it will be shown that the most varied sorts of external and subjective barriers account for the fact that capital accounting has arisen as a basic form of economic calculation only in the Western World.
>
> As distinguished from the calculations appropriate to a budgetary unit, the capital accounting and calculations of the market entrepreneur are oriented not to marginal utility, but to profitability. To be sure, the probabilities of profit are in the last analysis dependent on the income of consumption units and, through this, on the marginal utility structure of the disposable money incomes of the final consumers of consumption goods. As it is usually put, it depends on their "purchasing power" for the relevant commodities. But from a technical point of view, the accounting calculations of a profit-making enterprise and of a consumption unit differ as fundamentally as do the ends of want satisfaction and of profit-making which they serve. For purposes of economic theory, it is the marginal *consumer* who determines the direction of production. In actual fact, given the actual distribution of power, this is only true in a limited sense for the modern situation. To a large degree, even though the consumer has to be in a position to buy, his wants are "awakened" and "directed" by the entrepreneur.

> In a market economy every form of rational calculation, especially of capital accounting, is oriented to expectations of prices and their changes as they are determined by the conflicts of interests in bargaining and competition and the resolution of these conflicts. In profitability-accounting this is made particularly clear in that system of bookkeeping which is (up to now) the most highly developed one from a technical point of view, in the so-called double-entry bookkeeping. Through a system of individual accounts the fiction is here created that different departments within an enterprise, or individual accounts, conduct exchange operations with each other, thus permitting a check in the technically most perfect manner on the profitability of each individual step or measure.
>
> Weber, Max (1978): *Economy and Society. An Outline of Interpretive Sociology*. Berkeley: University of California Press, p. 92f.

While modern science aims at objectivity and impersonal validity the modern person discovers his or her individual subjectivity. Premodern cultures, too, could conceive of personhood and subjectivity, but these conceptions referred to "extraordinary" persons mainly to the ruler or the hero who created order instead of being subjected to it. One of the most striking and distinctive features of modern culture is that this creative personhood was extended to every citizen endowed with reason. Every individual person was seen as extraordinary, that is, as existing, partly at least, outside of the order of society. The individual citizens discovered their inner spaces of feelings and desires, and they were able to talk and write about it. Writing diaries became fashionable and the methodical conduct of life became a central concern, first with respect to the education of princes then also reaching out for ordinary citizens. Even before the Protestant reformation but strongly urged afterward, religion turned inward and focused on individual faith instead of ritual practices.

But this turn inward also engendered new problems. The gap between the inner feelings and the external behavior opened up. The question whether a spoken word or an action was a true and faithful representation of the speakers' inner intentions and feelings or whether it was just faked pretension, politeness, or a blunt lie took center stage. While premodern justice focused on the question of actual behavior, modern justice takes the intentions of the actor into account. Authenticity emerged as one of the key terms of modern morality. The literature of the eighteenth century, too, reflects this new turn. Theatrical performances centered masks and mistaken identities, stories about false priests and fake princes spread. Not every person, however, did already account for this difference between fiction and reality. Don Quixote took texts for granted reality and attacked windmills.

As science tried to describe nature as truthfully as possible, art tried to represent reality as accurately as possible. Since the fifteenth century the rendition of surfaces in pictorial art was increasingly refined, color and light became a central concern, the central perspective invented by renaissance art allowed to represent bodies in spaces as they are seen by the human eye. The illusion became a special artful mode of representation. The pictorial illusion of nature was perfected to such a degree that, from a distance, it was sometimes hard to decide whether something was real or painted. Looking at the ceilings of baroque palaces and cathedrals, the eye cannot decide any more whether it perceives a real column or only a painted one.

After the painting on canvas had replaced the al fresco painting on murals pictures could also be transported, collected, and traded. Princes and rich aristocrats assembled huge collections of pictorial art in their palaces thus representing important heroes, saints and ancestors, classical sites, and famous victories within their walls. Collecting art became a civilized form of trophy taking as had been collecting the relics of saints in medieval societies. The painter who was an esteemed craftsman before ascended to an icon of heightened subjectivity and personhood. The most famous of them—Michelangelo, Titian, and Rubens—were courted by kings, emperors, popes, and princes. They waited patiently to get their portraits commissioned, etc.

However, the rise of an illusionary art that tried to mirror the surfaces and the spatial nature of objects and bodies engendered, in a paradoxical turn, a crisis in the relationship to reality that was taken for granted before. If the power of illusion can be improved and if one rendition of surface and color can be more realistic than another, then pictorial representations as such do no longer mirror their real reference in a natural or evident way. If we strive for perfection we also assume that there are also misrepresentations and we know that the representation can, despite all our efforts, never fully grasp its reference. Thus the relationship between representation and represented reality appears as fragile and questionable.

A similar distrust in the illusionary character of representation can be noticed with respect to the scientific account of reality. Not all knowledge is true only because it claims so or because the authorities support it. Any claim to truth has to be critically investigated.

The incessant question whether a scientific statement is true or false changed also scholarly philosophy in Europe: while traditional philosophy searched for answers as to what eternal order was hidden behind the surface of phenomena, from the seventeenth century onward the question "how is true knowledge possible?" took centerstage. Metaphysics was replaced by the critical theory of knowledge.

Thus the once taken for granted relationship between representation and its reference or between signifier and signified was affected by questions and doubts. What, before, was a true symbol (e.g., the cross) that partakes in the sacredness of its reference (e.g., Christ), is now turned into a metaphor, an allegory or an imagination: we know that, by principle, all these metaphors, allegories, and imaginations can be replaced by others and that none of them can rightly claim to be the reality to which they refer. No law can claim to be God's will, no scientific truth can claim to be eternal, no behavior can claim to be absolutely authentic expression of inner feelings etc. This critical turn toward given representations is the message of the enlightenment.

Enlightenment, discovery, and publication were the order of the day in the early modern period. They assume that reality, truth, and the common good are obscured by private interests, secrecy, and everyday assumptions. In order to uplift the veil of tradition and ignorance the philosopher armed with suspicion and a critical attitude has to debunk hidden private interests and to turn the private to the public, to oppose the personal rule by impersonal moral principles and to reduce historical contingencies to the immutable laws of nature. Scientific discovery, moral education, and public critical reasoning were the prime tools to enhance personal perfection as well as the progress of humankind. Participating in this progress of humankind as a person endowed with reason was the new ideal of heightened subjectivity and personhood. Ascetic withdrawal from the pleasures of this world was superseded by the urge to belong to the avant-garde of history.

References

Castiglione, Baldassare (1900): *The Book of The Courtier. From the Italian of Count Castiglione: Done into English by Sir Thomas Hoby. Anno 1561. With an Introduction by Walter Raleigh.* London: David Nutt in the Strand.

Elias, Norbert (1978): *The Civilizing Process. The Development of Manners.* New York: Urizen Books.

Elias, Norbert (1984): *The Court Society.* New York: Pantheon Books.

Habermas, Jürgen (1989): *The Structural Transformation of the Public Sphere. An Inquiry into a Category of Bourgeois Society.* Cambridge: The MIT Press.

Weber, Max (1978): *Economy and Society. An Outline of Interpretive Sociology.* Berkeley: University of California Press.

Modern Nation States

Bernhard Giesen

Production: Industrial Revolution, Communication Revolution, Medical Revolution

In a way similar to the origin of conquest societies, the rise of the modern nation state, too, is based on breakthroughs in the relation between society and nature. The industrial revolution of the nineteenth century established a new mode of harnessing energy from nature and allowed for a hitherto unheard-of growth of productivity. Instead of using human or animal energy, industrial production was based on physical and, later on, on chemical energy transformed by steam engines, combustion motors and, later on, on electrical machines into movement. The new mechanical machinery consisted mostly of iron or steel that was produced by using coal and, later on, oil. Thus, in industrial societies, controlling resources of iron, coal, and oil increasingly took the position that the ownership of arable land had in premodern societies. In particular in the twentieth century, wars were fought for the control of these resources and wars could be fought only if these resources were in sufficient supply.

By contrast, the importance of agricultural production decreased—the use of mechanical energy and, later on, of chemical fertilizers allowed an ever-shrinking workforce to produce sufficient food for the entire population. While in the early modern societies more than two thirds of the total population were involved in agricultural production, the rural population of farmers, peasants, and their workforce went down to less than 20% in

B. Giesen et al., *Human Societies*, Cultural Sociology,
https://doi.org/10.1007/978-3-032-07697-7_8

advanced industrial societies at the end the twentieth century. In today's United States, the agricultural population figures at 3%. Furthermore, international trade and continuous exploitation of colonies shifted the production of food partly from the center of industrialism to the periphery of the modern world.

The part of the population that was set free from agricultural work left the rural areas for the cities to be employed as workers in the new industrial production sites or at least to offer their labor power. Thus, in the second half of the nineteenth century the urban population grew rapidly and the population of many European and American cities multiplied within a few decades. While London and Paris were large cities even before the start of industrialization, cities like New York or Berlin increased the number of their inhabitants by ten times within half a century.

Conditions of life in these large cities were often miserable—as it is in the large cities of the Third World today. Death from starvation was rare but due to the boast of the labor force many had to sell their bodies for bare survival. Children had to work in mines, while prostitution was an enormous business with many varieties. Street crime was a common problem.

> The bourgeoisie cannot exist without constantly revolutionizing the instruments of production, and thereby the relations of production, and with them the whole relations of society. Conservation of the old modes of production in unaltered form, was, on the contrary, the first condition of existence for all earlier industrial classes. Constant revolutionizing of production, uninterrupted disturbance of all social conditions, everlasting uncertainty and agitation distinguish the bourgeois epoch from all earlier ones. All fixed, fast-frozen relations, with their train of ancient and venerable prejudices and opinions are swept away, all new-formed ones become antiquated before they can ossify. All that is solid melts into air, all that is holy is profaned, and man is at last compelled to face with sober senses, his real conditions of life, and his relations with his kind.
>
> The need of a constantly expanding market for its products chases the bourgeoisie over the whole surface of the globe. It must nestle everywhere, settle everywhere, establish connexions everywhere.
>
> The bourgeoisie has through its exploitation of the world market given a cosmopolitan character to production and consumption in every country. To the great chagrin of Reactionists, it has drawn from under the feet of industry the national ground on which it stood. All old-established national industries have been destroyed or are daily being destroyed. They are

dislodged by new industries, whose introduction becomes a life and death question for all civilized nations, by industries that no longer work up indigenous raw material, but raw material drawn from the remotest zones; industries whose products are consumed, not only at home, but in every quarter of the globe. In place of the old wants, satisfied by the productions of the country, we find new wants, requiring for their satisfaction the products of distant lands and climes. In place of the old local and national seclusion and self-sufficiency, we have intercourse in every direction, universal interdependence of nations. And as in material, so also in intellectual production. The intellectual creations of individual nations become common property. National one-sidedness and narrow-mindedness become more and more impossible, and from the numerous national and local literatures, there arises a world literature.

The bourgeoisie, by the rapid improvement of all instruments of production, by the immensely facilitated means of communication, draws all, even the most barbarian, nations into civilization. The cheap prices of its commodities are the heavy artillery with which it batters down all Chinese walls, with which it forces 'the barbarians' intensely obstinate hatred of foreigners to capitulate. It compels all nations, on pain of extinction, to adopt the bourgeois mode of production; it compels them to introduce what it calls civilization into their midst, i.e., to become bourgeois themselves. In one word, it creates a world after its own image.

The bourgeoisie has subjected the country to the rule of the towns. It has created enormous cities, has greatly increased the urban population as compared with the rural, and has thus rescued a considerable part of the population from the idiocy of rural life. Just as it has made the country dependent on the towns, so it has made barbarian and semi-barbarian countries dependent on the civilized ones, nations of peasants on nations of bourgeois, the East on the West.

Marx, Karl and Friedrich Engels (2002): *The Communist Manifesto*. London: Penguin Books, pp. 222–224

The growth of the cities corresponds to a general increase of the population and a rise in average life expectancy. For centuries the population size and density of most European regions had been rather stable. At the end of the eighteenth century, France, the largest European nation, had 24, Britain 11, and Prussia 7 million inhabitants. Within one century Britain almost quadrupled its population, France went up to 40 million, Germany to 60 million.

After this steep increase during the nineteenth century, the total population remained relatively stable and grew only slowly. The main cause for this remarkable development is a continuation of traditional fertility patterns and the strong impact of modern medicine and hygiene. The number of children born in each family remained the same for several generations but the survival rates and the average life expectancy of newly born children increased considerably. Without modern medicine it would have been very low—in most states less than 25 years. People over 35 were regarded as old in premodern societies whereas, today, even retired people look forward to enjoying their life for many years. At the beginning of the population growth in the nineteenth century the demographic situation was very similar to the one that we encounter in many African or Latin American societies today: by far the largest part of the population is under 25, migration to the big cities is strong, unemployment rates in the urban centers are high. Whoever can afford it tries to escape from the misery at home and to get access to the promised land of new opportunities.

The rapid spread of mechanical production—first in Britain, then in continental Europe and in the United States, finally on the entire globe—is well known. It boasted an unheard-of growth of productivity and resulted in lower prices for goods and commodities. What had been individually crafted and slowly produced by manual labor before, could now be produced rapidly by large quantities. Still manual human labor was involved in the process of production, but it was split up into many standardized activities by the technological division of labor. This division of labor reduced human labor to filling the gap between different machines, and, finally, to directing the process, controlling its results and improving the technology required. The energy was generated by machinery.

> The greatest improvement in the productive powers of labour, and the greater part of the skill, dexterity, and judgment with which it is anywhere directed, or applied, seem to have been the effects of the division of labour. The effects of the division of labour, in the general business of society, will be more easily understood considering in what manner it operates in some particular manufactures. It is commonly supposed to be carried furthest in some very trifling ones; not perhaps that it really is carried further in them than in others of more importance: but in those trifling manufactures which are destined to supply the small wants of but a small number of people, the whole number of workmen must necessarily be small; and those employed in every different branch of the work can often be collected into the same workhouse, placed at once under the view of the spectator.
>
> In those great manufactures, on the contrary, which are destined to supply the great wants of the great body of the people, every different branch

of the work employs so great a number of workmen, that it is impossible to collect them all into the same workhouse. We can seldom see more, at one time, than those employed in one single branch. Though in such manufactures, therefore, the work may really be divided into a much greater number of parts, than in those of a more trifling nature, the division is not near so obvious, and has accordingly been much less observed.

To take an example, therefore, from a very trifling manufacture; but one in which the division of labour has been very often taken notice of, the trade of the pin-maker; a workman not educated to this business (which the division of labour has rendered a distinct trade), nor acquainted with the use of the machinery employed in it (to the invention of which the same division of labour has probably given occasion), could scarce, perhaps, with his utmost industry, make one pin in a day, and certainly could not make twenty. But in the way in which this business is now carried on, not only the whole work is a peculiar trade, but it is divided into a number of branches, of which the greater part are likewise peculiar trades. One man draws out the wire, another straights it, a third cuts it, a fourth points it, a fifth grinds it at the top for receiving the head; to make the head requires two or three distinct operations; to put it on, is a peculiar business, to whiten the pins is another; it is even a trade by itself to put them into the paper; and the important business of making a pin is, in this manner, divided into about eighteen distinct operations, which, in some manufactories, are all performed by distinct hands, though in others the same man will sometimes perform two or three of them. I have seen a small manufactory of this kind where ten men only were employed, and where some of them consequently performed two or three distinct operations. But though they were very poor, and therefore but indifferently accommodated with the necessary machinery, they could, when they exerted themselves, make among them about twelve pounds of pins in a day. There are in a pound upwards of four thousand pins of a middling size. Those ten persons, therefore, could make among them upwards of forty-eight thousand pins in a day. Each person, therefore, making a tenth part of forty-eight thousand pins, might be considered as making four thousand eight hundred pins in a day. But if they had all wrought separately and independently, and without any of them having been educated to this peculiar business, they certainly could not each of them have made twenty, perhaps not one pin in a day; that is, certainly, not the two hundred and fortieth, perhaps not the four thousand eight hundredth part of what they are at present capable of performing, in consequence of a proper division and combination of their different operations.

In every other art and manufacture, the effects of division of labour are similar to what they are in. this very trifling one; though, in many of them, the labour can neither be so much subdivided, nor reduced to so great a simplicity of operation. The division of labour, however, so far as it can be introduced, occasions, in every art, a proportionable increase of the productive powers of labour. The separation of different trades and employments from one another, seems to have taken place, in consequence of this advantage.

Smith, Adam (1976): *An Inquiry Into the Nature and Causes of The Wealth of Nations.* Chicago: University of Chicago Press, pp. 7–9

Industrial production shaped the character of the goods it produced. Individual craftsmanship was replaced by standardization. Whether it was nails, screws, stools, cars, canons, or computer screens, all industrially produced items existed and exist in large series of alikes that abstract from the individually changing needs of its potential users and presuppose an anonymous standardized consumer for standardized goods. Mass society is stamped by mass production. The lasting individual relationship between a person and her tools is turned into the quick acquisition of exchangeable tools by exchangeable users. Without this dissolution of the ties between persons and things and without this standardization of products it would, however, also be much more difficult to rely on the quality of products. If each tool in a series would differ from the others, predictability would decrease and time for learning and experience would increase. Recombination of parts would be seriously imperiled, comparison of prices on markets would be very difficult.

The use of mechanical and electrical energy allowed also for new modes of transportation and communication. Since the beginnings of human civilization and despite improvements in road construction, the speed of transportation and communication overland was mainly limited by the pace of horses. Now, for the first time, a major increase of this speed was possible: railroads and trains, later on, cars with combustion engines and aircraft reduced the time of traveling a distance overland to fractions of the one required by coaches and horses.

This acceleration of transportation was hailed with enthusiasm and its tools became icons of the modern motion across boundaries. The train, the car, and the aircraft replaced not only the horse on the level of physical transportation, but also as symbols of (mainly male) prestige and pride. On the other hand, however, the acceleration of transportation and communication was also provoking anxiety and fear when it was still new and uncommon: many were afraid of nausea and diseases caused by trains;

some expected natural countryside to be dissected by railroads and interpreted trains as vehicles of the devil.

The increase of speed by railroad transportation engendered also a new management of time. While the departure and arrival of ships could be expected only with regard to a specific day or, in case of shorter distances with respect to a broad time of the day (the "afternoon boat"), the movement of trains was scheduled by hour and minute. This resulted in a much more precise management of time and reduced the waste time of waiting. The mechanical clock started to reign the social world, punctuality replaced politeness as a major virtue and the coordination of social activities was increasingly shifted from authority to time.

Rapid railroad transportation was also the basis for quick and regular mail. Letters could be transmitted within days instead of weeks. The network of mail exchange became one of the core organizations of the modern nations. Later on, the invention of the telegraph and the telephone speeded up translocal communication to a degree that could hardly be imagined in premodern periods. Information and talk could travel across long distances without consuming time. This changed the imagination of social networks and devaluated the importance of local presence. At the end, in the world of mobile phones, we can talk everywhere whenever we want to whomever we want. Speaking to somebody is turned from a precious occasion that required some effort into a communicative nuisance, switching off the mobile and being inaccessible for others becomes a luxury—at least for some people.

Classification: Nations and Minorities, Functional Differentiation and Education, Individualism and the Quest for Community

The early modern territorial state had many features that strike us as being quite similar to contemporary modern societies: there was impersonal law and rational bureaucratic administration, the idea of progress and enlightenment, a centralized state and a standing army, fiscal politics and expanding markets, self-assured merchant capitalists and printed books, etc. But early modern society was still imagined and defined as a creation of the ruler—as enlightened and tolerant as his rule may have been. The monarch was subject to reason but sovereign with respect to the law. It was dynastic continuity and the range of monarchic authority that defined the state and ensured its identity. The term "society" referred still to the "good society" of respectable citizens and, in particular, to all those who were related to the princely court. The vertical stratification of class and honor still provided the backbone of this good society. Its salience was evident and did not need any justification.

This perspective from the center to the periphery, from above to below, from the sovereign to his subjects was challenged by the great *revolutions* that mark the transition to the modern nation state. The English, the American, and the French revolutions started as movements that were to restore the honest, good, and fair ways of life that were presumably spoilt or repressed by frivolous monarchs and decadent courtiers. They were not geared to create something new but, instead, to return to the good old days when decency reigned and the working citizen could have his fair share of the common good. And they were not triggered off by the insupportable poverty of the lower classes. Neither the Puritans who fought against the English king, nor the American colonists who rebelled against the British rule, nor the French notables who abolished the privileges of the nobility were poor and miserable outcasts who struggled against misery and tried to get rid of their chains. Instead, they were mostly respected, well-educated, and often wealthy members of the middle classes and they imagined themselves as such. Few of them thought about the liberation of slaves and servants or the equality of men and women. Instead, their concern was high taxes, the rights of citizens and the rule of virtue.

But in opposing the decadent and frivolous life of the princely court, and in standing up against the monarchic rule they also differed from previous resistance movements against higher taxes or centralized monarchic rule. They presented their own way of life as the superior morality and they strived not only to get access to the center, but they claimed to be the center of society. The famous question of the Abbé Sieyès, a member of the French national assembly, "What is the third estate?," and its assumed answer, "The third estate is the nation," marks this shift in the notion of sovereignty. The assembly of citizens conceived of themselves as the nation and this nation claimed to be the source of the law and the social order—thus reversing the relationship between the prince and his subjects. In the great revolutions this claim of the people to be the sovereign center of society took center stage. The people relapsed into a state of nature, seized violently the power, decapitated the personal embodiment of princely rule, and gave themselves a constitution. Inequality being the structural backbone of the princely state was replaced by the new claim of equality and fraternity of citizens.

This reversal from top to bottom triggered off a new movement of inclusion into the political community of the nation. At first it was just the male property owning and tax paying citizens who were seen as the core of the nation, later on also poor craftsmen, shopkeepers, servants, workers, women, finally the former slaves were included into the citizenry. This ever-expanding inclusion during the nineteenth century was, certainly the result of bitter struggles between those who defended their privileges and the old order and those who strived to accelerate the progress of history. But its success was backed by the axial reversal of the vertical order of rank and of the relationship between center and periphery. In modern society, the periphery and the bottom of society are seen as the source of creativity, agency, and authenticity whereas the center and the ruling class are under suspicion of being decadent and deceptive.

But this gradual inclusion into the nation of citizens could not dispense with the question of the boundaries of the nation in distinction of other nations. Who was to be excluded and on what grounds and reasons? At first it was the party lines in the revolutionary struggle that provided the criteria of membership: whoever fought for the revolutionary cause belonged to the nation, whoever opposed it represented the enemy of the people: Catholics, priests, and aristocrats were to be excluded, persecuted, and finally executed. What had started in the name of virtue, liberty, and independence precipitated soon in the abyss of killing those who, according to the guardians of revolutionary virtue, were "enemies of the nation." Their exclusion was increasingly disconnected from their personal activities as supporters or opponents of the revolutionary cause.

Disregarding their individual commitments, the enemies were soon turned into "aliens" who were excluded not because of their commitment but because they belonged to a certain category. The new democratic sovereign, the nation, presupposed homogeneity and continuity of this nation—in a certain respect at least. In the princely state it was the body and the will of the monarch that provided this homogeneity in an obvious way. Replacing the prince by the people, however, inevitably engenders the question as to on which basis the homogeneity of the nation is to be grounded and which categories of people are incompatible with this assumed homogeneity. The sheer coincidence of living on the territory of a state or the descent from parents who were recognized as citizens of the nation could increasingly be questioned as relying on criteria of the "ancien régime" that focused on descent and territory. Instead, a new prepolitical basis of the nation was required. Searching for these prepolitical foundations the nations imagined their homogeneity as resulting from nature, from climate, race, and ethnicity.

This ethnic idea of the nation seemed to solve a problem but it did so at the expense of causing an even larger one. The nation of co-ethnics was separated from minorities who lived on national territory but were defined as "aliens" who were to be expelled on ethnic reasons. What in early modern territorial states started as the requirement of confessional homogeneity of the citizens, was continued in the modern democratic nation state as the quest for ethnic purity and homogeneity. In the name of ethnic homogeneity modern nation states expelled people speaking a different language, praying to a different God, or claiming a different history. Since the majority rules in modern democracies, these people were called "minorities." The new structural principle of the nation state claimed identical boundaries for the people, the territory, the culture, and the law of the state. One language, one culture, one descent, one law, even one currency within the boundaries of the state. State controlled institutions—in particular compulsory education and military service, but also national academies and national museums—were to forge this unity of the nation. Their importance for the making of European nations in the nineteenth century can hardly be underrated.

Whoever did not fit into this identity model of the nation was accepted as a passing visitor and was granted permanent residence only in small numbers and exceptional cases. If, by contrast, the minority was a larger group of permanent residents who were rather visible in public spaces, then the majority sometimes reacted by violence, persecution, and ethnic cleansing. This repression of minorities occurred in rural as well as in urban areas. Although it was, in many cases, stimulated by populist political leaders, the question remains why the people responded so frequently to the seductions of the ethnic narrative. The inconvenient answer to this question is that ethnic cleansing is related to the cultural program of the modern nation state. In a way the "alien" minority is taking the position that the non-working vagabonds or the demonic monsters had in earlier societies—the person who does not fit into the grid of cultural classification, and who, therefore, has to be removed if he or she cannot be ignored any more. This position of the unclassifiable was, in particular, ascribed to persons who could not be regarded as passing visitors whose residence was in other nation states: Jews and Gypsies who had no homeland of their own became increasingly victims of ethnic exclusion and persecution.

> The unity of nearness and remoteness involved in every human relation is organized, in the phenomenon of the stranger, in a way which may be most briefly formulated by saying that in the relationship to him, distance means that he, who is close by, is far, and strangeness means that he, who also is far, is actually near. For, to be a stranger is naturally a very positive relation; it is a specific form of interaction. The inhabitants of Sirius are not really strangers to us, at least not in any sociologically relevant sense: they do not exist for us at all; they are beyond far and near. The stranger, like the poor and like sundry "inner enemies," is an element of the group itself. His position as a full-fledged member involves both being outside it and confronting it. The following statements, which are by no means intended as exhaustive, indicate how elements which increase distance and repel, in the relations of and with the stranger produce a pattern of coordination and consistent interaction.
>
> Throughout the history of economics the stranger every-where appears as the trader, or the trader as stranger. As long as economy is essentially self-sufficient, or products are exchanged within a spatially narrow group, it needs no middle-man: a trader is only required for products that originate outside the group. Insofar as members do not leave the circle in order to buy these necessities—in which case *they* are the "strange" merchants in that outside territory—the trader *must* be a stranger, since nobody else has a chance to make a living.
>
> This position of the stranger stands out more sharply if he settles down in the place of his activity, instead of leaving it again: in innumerable cases even this is possible only if he can live by intermediate trade. Once an economy is somehow closed, the land is divided up, and handicrafts are established that satisfy the demand for them, the trader, too, can find his existence. For in trade, which alone makes possible unlimited combinations, intelligence always finds expansions and new territories, an achievement which is very difficult to attain for the original producer with his lesser mobility and his dependence upon a circle of customers that can be increased only slowly. Trade can always absorb more people than primary production; it is, therefore, the sphere indicated for the stranger, who intrudes as a supernumerary, so to speak, into a group in which the economic positions are actually occupied—the classical example is the history of European Jews. The stranger is by nature no "owner of soil"—soil not only in the physical, but also in the figurative sense of a life-substance which is fixed, if not in a point in space, at least in an ideal point of the social environment. Although in more intimate relations, he may develop all kinds of charm and significance, as long

> as he is considered a stranger in the eyes of the other, he is not an "owner of soil." Restriction to intermediary trade, and often (as though sublimated from it) to pure finance, gives him the specific character of *mobility*. If mobility takes place within a closed group, it embodies that synthesis of nearness and distance which constitutes the formal position of the stranger. For, the fundamentally mobile person comes in contact, at one time or another, with every individual, but is not organically connected, through established ties of kinship, locality, and occupation, with any single one.
>
> Simmel, Georg (1950): "The Stranger". In: Wolff, Kurt H. (ed.): *The Sociology of Georg Simmel*. Glencoe: The Free Press, 402–408, pp. 402–404

Pogroms against Jews were not an invention of the modern nation state. But the new classificatory model of national purity fostered the resonance of anti-Semitism even among those educated classes who had despised and opposed anti-Jewish violence before.

But the nation state is not inescapably linked to the idea of ethnic purity. Some early modern democracies like Switzerland can even dispense with the idea of a common language. These relatively stable multiethnic nation states originated mostly in a revolutionary upheaval against foreign occupation or princely repression and grounded their national identity on the public memory of this historical act and on the commitment to certain national virtues and values. Instead of being rooted in the prepolitical nature of every single citizen, the nation is here constituted by participation in public life, by commitment to traditions and values. Whoever is able to join this commitment can be included into the citizenry—disregarding descent and origin. This opens, in principle, also the boundaries to immigration for those who are willing to assimilate to the national culture and to respect the law of the state. It is no coincidence that the United States cutting the ties between citizenship and descent were, for most of its history, not only open to immigration but depending on it.

But even this idea of the nation as a community of values and traditions engenders certain risks. First it might be objected that the United States with its missionary zeal and its global reach out comes closer to the structure of empires than to the ones of nation states (Switzerland too might be treated rather as a confederation with a very weak center instead of being a nation state). More important, however, is the occasionally strong pressure to close the borders to massive immigration that, according to many citizens, is threatening their economic welfare. Furthermore, the existence of large homogeneous communities of immigrants on the territory of the nation state does often preclude their assimilation to the national culture and their participation in the public sphere; within their communities there are few pressures to learn the language of the new homeland, lack of language skills results in suboptimal school scores and high unemployment, poverty and failure of economic integration drive many into criminal subcultures—segregation increases.

While nationality was the new mode of constructing the unity of society its internal structure was increasingly patterned by functional differentiation. Functional differentiation was preluded by the division of labor in premodern societies. This division of labor did not consist only of separating different technical procedures and activities but it institutionalized this separation by establishing professional competences and, sometimes even monopolies in performing certain tasks, producing certain goods and delivering certain services. Only an examined and certified member of a professional association was entitled to offer this specialized labor in due and decent ways, at fixed prices, and with a warranty of the association. This rigidified division of labor with its lack of competition was increasingly dissolved in modern societies but the idea of a differentiation of functions prevailed and ascended to the position of a core classification of modern society.

Each field of action was defined by its own autonomous rationality and had to abstract from every other regards. In the field of economic activities it is the expected relationship between costs and benefits that is exclusively to determine choices and decisions. An entrepreneur who continuously ignores this economic rationality and sells below market prices because he respected personal relationships or traditional affiliations would soon be out of business. He can, at best, claim to be a charitable individual. Markets enforce their own rules by a non-violent way of sanctioning. Whoever disregards the rules slips out of the field of economic behavior. This merciless rule of impersonal codes does not only exclude those who do not act according to it, but, by its very impersonal validity, it allows also to include every person who acts accordingly. Markets are expansive and inclusive.

Costs and benefits are however, of no concern in the field of jurisdiction. Decisions in court have to disregard cost and focus exclusively on the law. Other references have likewise to be excluded. Kinship or love relationships between the judge and the accused persons in a penal court have to be prevented, the beauty and the manners of the accused person do not matter. Jurisdiction has to be performed according to the law and to nothing else. Here again, the exclusion of personal ties engenders the autonomy of the modern jurisdiction and the inclusion of every person who obeys its rules.

Political authority has to be cut from descent and is to be based on the will of the nation instead of military repression of the people. This will can be expressed by regular elections or by other institutional devices that allow to map changing consent.

Religion and politics are to be institutionally disconnected. Membership in a particular religious community does not impact citizenship and vice versa. This autonomy of religious confession and the possible coexistence of several religions within one political community is at the core of the modern nation state. It allows communities to overcome internal confessional wars and to include people into the citizenry regardless of their confession.

A similar autonomy is claimed for science and art, and even for personal relationships such as marriage. Since the beginnings of modernity the decision to marry a person is to be based on personal love and sympathy alone and other reasons are largely excluded—at least from public justification. Whoever is pressed to marry for reasons of power or money is regarded as an object of pity or as a figure of tragic stories.

Scientific research is to be moved by the quest for truth and can dispense with the question of justice and beauty. A scientific report can hardly be challenged with respect to its clumsy style. Literature, by contrast, does not have to respect truth and justice, but is bound to realize aesthetic values, such as novelty or artful reference to the classics of the genre.

> 'Scientific' pleading is meaningless in principle because the various value spheres of the world stand in irreconcilable conflict with each other. The elder Mill, whose philosophy I will not praise otherwise, was on this point right when he said: If one proceeds from pure experience, one arrives at polytheism. This is shallow in formulation and sounds paradoxical, and yet there is truth in it. If anything, we realize again today that something can be sacred not only in spite of its not being beautiful, but rather because and in so far as it is not beautiful. You will find this documented in the fifty-third chapter of the book of Isaiah and in the twenty-first Psalm. And, since Nietzsche, we realize that something can be beautiful, not only in spite of the aspect in which it is not good, but rather in that very aspect. You will find this expressed earlier in the *Fleurs du mal*, as Baudelaire named his volume of poems. It is commonplace to observe that something may be true although it is not beautiful and not holy and not good. Indeed it may be true in precisely those aspects. But all these are only the most elementary

> cases of the struggle that the gods of the various orders and values are engaged in. I do not know how one might wish to decide 'scientifically' the value of French and German culture; for here, too, different gods struggle with one another, now and for all times to come.
>
> We live as did the ancients when their world was not yet disenchanted of its gods and demons, only we live in a different sense. As Hellenic man at times sacrificed to Aphrodite and at other times to Apollo, and, above all, as everybody sacrificed to the gods of his city, so do we still nowadays, only the bearing of man has been disenchanted and denuded of its mystical but inwardly genuine plasticity. Fate, and certainly not 'science,' holds sway over these gods and their struggles. One can only understand what the godhead is for the one order or for the other, or better, what godhead is in the one or in the other order. With this understanding, however, the matter has reached its limit so far as it can be discussed in a lecture-room and by a professor. Yet the great and vital problem that is contained therein is, of course, very far from being concluded. But forces other than university chairs have their say in this matter.
>
> Weber, Max (1967): "Science as a Vocation". In: Gerth, Hans H. and C. Wright Mills (eds.): *From Max Weber: Essays in Sociology.* New York: Oxford University Press, p. 147f.

In this way every domain is subject to its own autonomous rules. Any attempt to cross the boundaries of these domains and to violate this autonomy can be challenged and scandalized as corruption. Kinship ties and money are not to influence the decisions in public office, political power must not interfere with scientific truth, public office is not to generate private fortunes, etc. The moral of functional differentiation consists of respecting the boundaries of autonomous spheres. Whatever the hidden motives of decisions may be, they can be publicly justified only with respect to the special rules and codes of the respective domain.

This separation of different functional domains and the ban on corruption seem to be so obvious and evident to us that we can hardly imagine a society whose institutions could operate effectively without respecting these rules. But for most of human history the institutional differentiation of domains was rather weak and this porosity of boundaries even had some advantages: it prevented problems of coordination. Political authority could generate economic resources and vice versa, religious leadership coincided with political authority, kinship ties defined duties and justice, etc.

The cultural classification of the modern nation state does not only consist of the nation as the conception of its unity and functional differentiation as its internal system of boundaries. It did not erase family ties or social hierarchies, male associations or the tension between center and periphery. All these modes of classification continue to exist but they are superseded by the new ideas of functional differentiation and national sovereignty. This salience of functional differentiation is mirrored by a change in the prime dimension of social classification. While age and descent, wealth and power still matter for classifying individuals, the most important basis for social identity is shifted to the realm of work and occupation. In modern nation states it is your occupation and the kind of work you do for a living that decides about social standing, prestige, and even style of life. Occupation and profession are no longer determined by birth and descent, but they are also not just a matter of choice and competition. Instead, they are largely determined by differences in education. All citizens in modern nation states are included into a compulsory elementary education that provides basic skills, mostly reading and writing. Higher education extends only to a part of the total population and academic training at universities is reserved for those who passed the institutional selection for talent and can afford the expenses. Thus social classification in modern nation states merges educational hierarchy with the division of labor between different professions or occupations. Among those who had a similar level of education functional differentiation determines social position. For example, medical doctors, lawyers, scientists, and corporate managers provide different professionals services but can meet on an equal footing.

This tendency to question vertical social inequality and to dissolve social boundaries derived from birth and descent in modern societies engenders a fluidification of social classification and shifts attention to the individual person who is regarded as the creative source of history and society. In modern societies the individual person, her dignity, her creativity, and her unalienable rights become the sacred center of society.

> The human person, whose definition serves as the touchstone according to which good must be distinguished from evil, is considered as sacred, in what one might call the ritual sense of the word. It has something of that transcendental majesty which the churches of all times have given to their Gods. It is conceived as being invested with that mysterious property which creates an empty space around holy objects, which keeps them away from profane

contacts and which draws them away from ordinary life. And it is exactly this feature which induces the respect of which it is the object. Whoever makes an attempt on a man's life, on a man's liberty, on a man's honour inspires us with a feeling of horror, in every way analogous to that which the believer experiences when he sees his idol profaned. Such a morality is therefore not simply a hygienic discipline or a wise principle of economy. It is a religion of which man is, at the same time, both believer and God.

But this religion is individualistic, since it has man as its object, and since man is, by definition, an individual. Indeed there is no system whose individualism is more uncompromising. Nowhere are the rights of man affirmed more energetically, since the individual is here placed on the level of sacrosanct objects; nowhere is he more jealously protected from external encroachments, whatever their source. The doctrine of utility can easily accept all kinds of compromises, without denying its fundamental axiom; it can allow that individual liberties should be suspended whenever the interest of the greatest number demands this sacrifice. But there is no possible compromise with a principle which is thus put above and beyond all temporal interests. [...]

Not only is individualism distinct from anarchy; but it is henceforth the only system of beliefs which can ensure the moral unity of the country. [...]

Now, all the evidence points to the conclusion that the only possible candidate is precisely this religion of humanity whose rational expression is the individualist morality. To what, after all, should collective sentiments be directed in future? As societies become more voluminous and spread over vaster territories, their traditions and practices, in order to adapt to the diversity of situations and constantly changing circumstances, are compelled to maintain a state of plasticity and instability which no longer offers adequate resistance to individual variations. These latter, being less well contained, develop more freely and multiply in number; that is, everyone increasingly follows his own path. At the same time, as a consequence of a more advanced division of labour, each mind finds itself directed towards a different point of the horizon, reflects a different aspect of the world and, as a result, the contents of men's minds differ from one subject to another. One is thus gradually proceeding towards a state of affairs, now almost attained, in which the members of a single social group will no longer have anything in common other than their humanity, that is, the characteristics which constitute the human person in general. This idea of the human person, given different emphases in accordance with the diversity of national temperaments, is therefore the sole idea that survives, immutable and impersonal, above the changing tides of particular opinions; and the sentiments which it awakens are the only ones to be found in almost all hearts.

Durkheim, Émile (1969): "Individualism and the Intellectuals". In: Lukes, Steven: "Durkheim's 'Individualism and the Intellectuals'". In: *Political Studies. Volume 17(1)*, 19–30, pp. 21f., 25f.

The distinction between individual personhood and individual liberties on the one hand and social positions and obligations on the other was already a central concern of the social elites of the eighteenth century. In modern society it is, however, extended to each and every citizen. Every adult individual citizen has not only the right to participate in decisions about politics, to choose, according to his or her means, between alternative offers on markets, to move freely on the territory of the state, and even to leave the state and to seek fortunes abroad, but he or she has the obligation to define and to determine his or her unique personhood by herself. Every modern individual is to become the creative artist of his or her own personality. And it is not by coincidence that, in the nineteenth century, the figure of the artist ascends to an icon of heightened personhood and subjectivity thus replacing the figure of the ascetic or progressive individual.

But this general claim for the autonomy of the individual person is not without risks. What had been the liberation from the external oppression before is, after its completion, turned into the burden of inventing yourself in distinction to others. Looking into the empty inner spaces of our selves can be a frightening experience. Most people in modern societies try to ignore this burden of autonomy and imitate those who seem to be able to accomplish this task: charismatic public figures like artists and pop stars, people who create fashion and style, claim virtues and values, and provide orientation and distinction. The first who follow the new trend can still see themselves as the avant-garde in distinction to the many others, but, since there are no social barriers to imitation and copying, many of these others can join the avant-garde soon, thus turning the unique mark of individuality and distinction into the common property of everyone. At the end the once adored charismatic figure who had started the "original" new trend is discovered as a boring oddity, as a shallow pretender, as outdated, as a fake. This devaluation of creative individualization to its banal opposite is the inevitable result of the fluidification of social boundaries and social ties in modern societies.

If the turnover of fashion and the dynamics of devaluation accelerate a new yearning for stable roots and binding ties comes up. The immutable embeddedness into a community that premodern people tried to overcome becomes now the goal of a new quest. This new quest for certainties leads to a range of social movements, among them a rediscovery of tradition and family, a search for the lost harmony between nature and society, a revival of ethnicity and locality, and diverse kinds of fundamentalist commitments strive to overcome alienation and decadence. These movements differ profoundly from the class-centered movements that aimed at overcoming poverty and civic exclusion, social inequality and disenfranchisement. While members in these class-centered movements act on their own behalf and try to improve their own conditions of life, the new social movements advocate a cause that refers to the right social order or to the collective identity of a community.

Circulation: Public Media Communication, Democratic Elections, Financial Markets

Many may consider the modern welfare state to be the most important new institution of circulation. Indeed the change from kinship solidarity and private charity to a massive redistribution based on a multitude of legal entitlements can rightly be regarded as a fundamental breakthrough in supporting not only the poor, but also all those who are unable to support themselves: handicapped and elderly people, physically disabled and jobless, students, etc. While modern nation states vary vastly with respect to the extent of welfare transfers there is no nation state that is completely devoid of it. In some Scandinavian states the legally entitled welfare transfers used to be a much more important source of revenue for the population than salaries or earnings from labor. In the United States, by contrast, state-based welfare transfers are only a very small part of the gross national product. The pervasive existence of welfare redistribution in modern societies is related to the very notion of the nation as the basic community. Whoever has citizenship rights can claim the solidarity of his co-nationals in case she is impaired to earn her living by work. The same logic of national solidarity applies to progressive taxation for citizens with higher incomes. Without the underlying assumption of communal obligations people with higher incomes would hardly accept higher taxes as a legitimate contribution to the common wealth—since most of them do not make more use of public goods (roads, schools, etc.) than people who are less well off.

After this short reference to the centralized redistribution system of welfare states we will outline three new institutional systems by which information, power, and property are circulated in modern societies; public media of communication, democratic elections, and financial markets. They converge in several respects: (1) they focus on competition, alternative options, and polycentrism; (2) they have open and porous boundaries and try to include as many people as possible; (3) they rely on an institutionalized procedure of communication that accelerates circulation (of news, political leaders, property rights); and (4) they map the complexity of their respective environments (public audience, voters, investors) in a finely tuned way.

From the eighteenth century onward the early modern territorial state had a flourishing public sphere based on associations, salons, and printed newspapers. But the participation in public communication was limited to those who were able to read and to write, that is, to a relatively small part of the population. The egalitarian program of the nation state, however, aimed at the inclusion of its entire citizenry into the public sphere. This could be achieved only on the basis of literacy which, on its turn, required mandatory schooling for the entire population. Mandatory schooling and compulsory military service provided the institutional instruments of the nation state for drawing young people out of their families and including them into the encompassing community of the nation.

The expansion of the public sphere in modern nation states came, however, at a price. While the public sphere of the eighteenth century provided a space where, in principle, every educated citizen could address an anonymous community with respect to common issues, the expanded public sphere of the modern nation state had to limit access to authorship. Very few individuals are writing for a large readership and these writers are mostly professional journalists who compete for the attention of the public audience.

This changes the rules of public communication. Although most journalists pretend to write out of concern for the common good of the nation, most of them strive for catching the attention of a large public audience. News has to sell. And news is only news if it is different from yesterday's news. The modern public sphere is turned from a civil arena into a competitive business. Since the foundation of journals and, later on, radio and TV broadcasting depends on the size of the audience, the message has to please the common taste and it has to be sensitive to changes. Lengthy and demanding deliberations tend to be replaced by easy entertainment and sensational reports that appeal to everybody and require little intellectual efforts. Images and movies gain in importance, the message has to be transmitted in a few moments, headlines and faces decide about reading the article or staying with a TV channel instead of switching attention to the next one. Nothing remains secret and private, everything can become the issue of public attention, an ever-expanding galaxy of public entertainment seems to suffocate public reasoning about common issues.

But despite all short-lived entertainment and the staging of human touch (pets, kids, crimes, and the misery of others), modern public spheres contain still spaces of serious discourse about the common good, about politics, and about general issues of the human condition. Only a small part of the total population participates in this discourse but those who do are among the more influential individuals in modern societies.

Since the nation is the sovereign in the modern nation state, legislation and government have to be based on the will of the people. Modern nation states are thus representative democracies in the widest sense of the term. Political authority has to assume and to claim that it represents the will of the people. This claim may, however, interfere with the long-term perspectives on the common good or the day-to-day exigencies of political business. Politics has, therefore, to be partly decoupled from the explicit day to day consent of the people. Most modern nation states use regular elections to provide legitimacy and to allow for changing political leaders, but provide also for a certain independence of politics from the consent of the people for the period between the elections. Furthermore, in most cases, the citizens vote for representatives who carry on the daily political business on their behalf and, in many cases, also vote for the head of government. Thus, electoral periods and parliamentary representation provide institutional "breakers" between the will of the people and the political authority.

No viable procedure of election in modern societies can avoid restricting the political options between which the voters can choose, but democracies vary considerably with respect to the range of these options and the threshold to be passed for entering the political arena. They differ also with respect to the extent to which political opposition can be voiced and organized in the public sphere. In many modern nation states running for political office in an electoral campaign requires huge funds that are provided by parties, by donors, or by the personal fortune of the candidates. This restricts the number of candidates in a similar way as professional journalism limits the number of authors who get access to the public sphere. Some democracies claim to represent the will of the people by only one party that allows at best a yes or no vote, others are based on the institutionalized conflict between two parties upon which the voters can decide, and again others offer a vast range of different parties to choose between. At the first glance these multiparty systems seem to be more open and sensitive for articulating the will of the people, but they can, in most cases, arrive at majority decisions only by coalitions between several parties and this inevitable compromise distorts the original intentions and expectations of the voters. Thus multiparty systems that offer a variety of choices for voters invite voters to engage in political life but are bound to disappoint them after the elections. By contrast, one- or two-party systems are largely insensitive to the variation of the voters' intentions and do not invite to participate in public politics. Turnout by elections is, therefore, mostly low if there is no legal obligation to vote. Politicians who get electoral support in these systems do not have to compromise their agenda. Postelection disappointment is therefore lower.

Elections are decided by majority rules: whoever wins the majority of the votes of the people or the support of the majority of representatives is able to claim political authority. It has been objected that a small margin of several percents that ensures victory in democratic elections does not imply the true support of the people. Possibly the voters of the defeated candidate supported her very strongly and rallied for her in public whereas the winner might have got a lukewarm support from a tiny majority. Elections disregard the strength of the support: one person one vote. Throwing dices would be not more irrational than relying on marginal majorities. But for modern democratic institutions the question as to whether "the true will of the people" is accurately represented by the outcome is far less crucial than the respect of the institutionalized rules. Challenging the constitutional procedures for whatever good reasons jeopardizes seriously the modern political order which is rather based on procedural rules than on personal authority or on convictions. In a similar way we can question whether a piece of money seriously represents the value it claims or whether it is basically useless. However valid these doubts may be, we have to disregard them if we do not want to risk massive inflation. Money can mediate economic exchange only if we trust in its value. Or to use an illustration from technology: introducing a new standard of connectors may yield superior results, but it will be incompatible with all the existing connectors. Therefore established standards are rarely changed even if superior alternatives are available. A similar logic of procedural rationality holds true for democratic elections. Electoral institutions cannot avoid flaws, distortions of the will of the people and counterfactual assumptions (e.g., the autonomy of each single voter), but they have to be respected because they are the procedural core of democracies. They can claim an almost sacred position in the modern political system. Attempts to disregard them amount to suspending the political order and to risk the Hobbesian war of all against all.

Electoral procedures are not limited to the realm of politics. Public opinion, research, and marketing use similar procedures in order to monitor changes in the people's preferences, evaluations, and expectations. Here, too, the will of the people is not naturally given and obvious, but it has to be explored and visualized by special procedures and devices. And here, too, doubts may be raised and questions may be asked as to whether these procedures deliver an accurate picture of the true will of the people. But changes can only be assessed if the instruments of exploration are not altered. Therefore keeping a flawed procedure is sometimes more important than exchanging it by an improved one.

A new mode of circulation is also established by financial markets. Money and markets, even interests rates for money loans and the social separation between capital owners and workers existed already before the rise of the modern nation state. In contrast to the coercive labor prevalent in premodern societies the workers in modern societies were free to sell their workforce on labor markets to the owners of capital who invested in machinery, enterprise, and labor for profit.

At the beginnings of capitalism the owner of the enterprise was also its manager and its major source for technological innovation. This original unity of ownership, management and engineering was increasingly dissolved in the late nineteenth and in the twentieth century. The split between management and ownership is brought out most strikingly in the institution of financial markets.

Financial markets monitor the value of bonds, currencies, or property shares of corporations in a finely graded way. Instead of a single owner who acquires a corporation for a certain price and then controls and uses it on his own behalf, the property of a corporation is split into many shares that can be bought and sold on the stock market by many partitioned investors. The price of the stock or bond varies according to the expectations of future gains or losses. As democratic elections represent the will of the nation only in a distorted way, the price of the stock on financial markets does not mirror the actual worth of the corporation, the bond, or the resource. Instead, it monitors relative changes in expectations and comparative advantages ofinvesting in a certain stock or bond at a given moment.

Financial markets separate the position and interests of shareholders strictly from the internal control and direction of a corporation. They are not driven by the effort to save a particular corporation or to serve the interests of workers of this corporation but by the exclusive motive to buy what promises future gains and to sell what is likely to lose.

Since the future movement of stock is highly contingent, financial markets are very sensitive to rumors and insider information. Political changes and assumed trends impact heavily on them. Because there is no rational or methodical procedure to predict stock movements the business of professional investing in financial markets requires a spirit of risk taking and adventure that corresponds to the mind of warrior heroes in horticultural societies.

Financial markets presuppose fast comparisons of offers and rapid transactions. This fast pace is achieved by limiting the market exchange to a few hours a day and by confining it to a special locality where investors or their representatives meet and communicate with each other. The communication on financial markets is the opposite of bazaar exchange: there is no appraisal of the wares, no bargaining, no lengthy deliberation—just prices and the confirmation of the deal.

Representation: Autonomous Art and the Mass Media

Modern societies are patterned by functional differentiation establishing autonomous fields with logics of their own. This also extends to art. Early modern representative art referred still to the depiction of nature and aimed at a perfect illusion. This reference is increasingly rejected by modern art. Instead of striving for the perfect illusion of nature one branch of modern art tries to represent something that escapes from being represented by regular means: emotions, dreams, fantasies, insanities, the demonic nature of things. This reference to the hidden world behind the surface of things runs from romanticism to expressionism and surrealism. The taken for granted world of everyday life as well as the collectively shared grand symbols of beauty are shunned as "kitsch" by this art that plunges into the abyss of subjectivity and tries to debunk what is commonly ignored in social communication. Modern art has to shock and to surprise, it has to show the unknown, it has to make a difference to the familiar world of images. Art is art not because of the craftsmanship involved in its production but because it imposes a new perspective on the world. Few works of art show this programmatic commitment better than Magritte's famous painting that depicts a pipe with the statement "This is not a pipe" written on it or the so called "Ready Mades" (Marcel Duchamp), that is, ordinary objects of everyday life put on a pedestal in a museum. Modern autonomous art declines references to practical utility, conventional beauty, or political correctness. It is ambiguous, undecided and puzzling. It refers to itself and opens up a view to things before language and society can classify them.

This radical reference to itself is brought out very clearly by abstract art in the twentieth century. Abstract art has removed any reference to the world outside of art. It focuses on color and form and does not represent anything but itself. In its most radical vanguard version modern art claims that a painting is a just canvas covered with paint. At the beginning of the twentieth century cubism started to deconstruct the representation of everyday objects by rearranging their basic forms and by inserting pieces of newspaper on the surface of the painting. The newspaper was not to be a carrier of a message but it was to be regarded as an esthetical object. Viewers of these paintings had to give up their routinized ways of coping with a newspaper, that is, by reading it.

Later on, in minimalist "all over" paintings, any form was banned from the painting. The difference between background and sign inscribed on it disappears in these paintings. At the end, it was just white or black surfaces or geometrical sculptures—pure form or pure emptiness.

Another mode of self-reference consists of citing other pieces of art whether from popular or high culture. Pop artists like Warhol or Lichtenstein used well known aesthetic icons and transformed them into objects of modern art by blowing up their size or by arranging them in a repetitive way. Again modern art rejects any foundation in craftsmanship or in the significance of its content. Art is an autonomous perspective focused exclusively on aesthetics as science is an autonomous perspective focused on the discovery of truth.

While modern art disregards and occasionally even disdains easy acceptance by a large audience, the new mass media are set up for this broad acceptance. The following remarks will, however, not focus the new ways of circulating news in the public sphere of modern societies (see previous chapter) but deal with social effects of the mass reproduction of visual and aural representations in modern societies. Technical innovations as the printing of etchings, photographic recording, cinematography, audio-recording, radio, and television allowed for the uniform dispersion of images and music among a huge audience. Attending a theatrical or musical performance or looking at a painting requires presence at the local site of the event or at the place where the image is on display. People had to move or to travel, they had to get admitted, enjoying a pictorial, theatrical, or musical representation required efforts, presupposed a certain connoisseurship, and was often a matter of privilege. The performance was an event that attracted people to a certain location, focused their attention, and was talked about. Only very wealthy individuals or corporations such as churches or city councils could afford to sponsor these performances or to own a collection of paintings. This exclusivity of performative or pictorial representation is overcome by the mass reproduction of images and music in modern society (Benjamin 1968). Today we are able to enjoy music wherever we are, at home or in subway, in the gym or in the supermarket; our everyday life is flooded with images that compete for our attention; we can watch TV in the kitchen and in the bathroom; commercial advertising tries to catch our gaze and to invade our computer screens; images, music, and short pieces of texts are blended into a fireworks of representation in which nothing remains long and nothing requires the toils of understanding a difficult message. Today watching recorded images and listening to musical tunes are hardly precious, exclusive, and extraordinary events but they are part of our everyday environment and their inflationary spread tends to devaluate their importance. Nobody focuses her attention on the permanent drizzle of easy music in supermarkets or on TV commercials when the TV is never shut down. Images and sounds are like wallpapers of modern life.

The easy access to the media and their omnipresence has a close affinity to the inclusive and egalitarian structure of modern societies. As viewers of the TV program we are all equal and the cost of participation in the flow of images, sounds, and texts is rather low (many households below the poverty line can, however, not even afford a TV). Furthermore the multiplicity of alternative TV programs, movies, and musical recordings corresponds to the idea of the modern self who as a consumer, voter, or investor can choose between different alternatives. But, in seemingly paradoxical turn, this easy access and this broad range of alternatives have also increased the isolation of modern individuals. Watching TV at home or listening to music on headphones in public spaces does not connect people as a theatrical performance or a walk on the main square of a town in the nineteenth century did. We are as consumers of TV programs, CDs or DVDs alone with the music, the movies or the talk shows we opted for and, in most cases, we do not know what the others are watching or listening to. Thus the spread of modern mass media engenders a decline of a common public sphere.

The modern mass media altered the world of images as printing altered the world of texts. After centuries of the "Gutenberg galaxies," in which the written and printed text was the most important mode of symbolic representation, the position of written or oral text seems to be increasingly challenged by the rise of images. The art of oral speech is in decline since several decades. Few political leaders are still able to deliver a speech without a written manuscript. PowerPoint presentations abstract from complex subjunctive modalities and reduce the text to its bare essentials. The blending of short texts, background sound, and—mostly photographic—images becomes the most convincing mode of media communication. The convincing power of photographic images is due to their assumed "realism." Although we know how easy photo or video images can be manipulated, they move us much more than a written description. In contrast to texts, it is difficult to argue with them, to question them, or to debunk them as fake, fraud, or fantasies. Photographic images have a stronger "presence" than texts. In a certain way they are taken more for traces of an absent reality than for images: the existence of the absent reality, however distorted by the photo, can hardly be denied. This implicit claim of realism is supported by the tendency of the photographer or moviemaker to recede much more behind the representation than the author of a text or the artists of a painting.

The modern world of images may be devoid of the modifications of validity that texts dispose of (questioning, negating, commanding, describing, etc.). It is however patterned by iterations, citations, references, and genres that the creator of images and the viewers are familiar with. Thus the image is never just a mirror of reality but instead a visual representation of reality that repeats and hints at other images, icons, or common styles available in the repertory of representation. This holds true for all modes of cultural representation hence also for photos or movies.

The accelerated pace of circulation in modern societies engenders an increased mobility of people, resources, and information. Individuals change their location, their resources and their attitudes and these moves can be open or covered. Mapping the structure of society and determining an individual's position tend to be more difficult challenges in modern societies than premodern ones in which citizens were closely tied to localities and social groups. This results in new modes of representation. The private life of individual citizens has to be investigated and revealed to the public eye. Special institutions such as criminal investigation or public opinion polling aim at crossing the boundary between the hidden private life and the public knowledge of it. Identifying the unknown individual characteristics by figures and stories in the public sphere becomes a new field of representation. Criminal investigations in modern societies are less about finding whether an accused person had committed the crime than about identifying the criminal among a multitude of innocent citizens. Public opinion polling would not be necessary if the attitudes, preferences, and expectations of individuals would be determined by origin and descent, class position and occupation once and for all. The claim for autonomy of the individual person questions and dissolves these ties in modern nation states—to a certain degree at least. Since the private and the public, the individual person and the social order are treated as separate realms the one has to be represented in the other.

References

Benjamin, Walter (1968): "The Work of Art in the Age of Mechanical Reproduction". In: *Illuminations. Essays and Reflection.* New York: Shocken Books, 217–251.

Durkheim, Émile (1969): "Individualism and the Intellectuals". In: Lukes, Steven: "Durkheim's 'Individualism and the Intellectuals'". In: *Political Studies. Volume 17(1)*, 19–30.

Marx, Karl and Friedrich Engels (2002): *The Communist Manifesto.* London: Penguin Books.

Simmel, Georg (1950): "The Stranger". In: Wolff, Kurt H. (ed.): *The Sociology of Georg Simmel.* Glencoe: The Free Press, 402–408.

Smith, Adam (1976): *An Inquiry Into the Nature and Causes of The Wealth of Nations.* Chicago: University of Chicago Press.

Weber, Max (1967): "Science as a Vocation". In: Gerth, Hans H. and C. Wright Mills (eds.): *From Max Weber: Essays in Sociology.* New York: Oxford University Press, 129–156.

Postscript: Global Regimes

Bernhard Giesen

After the collapse of the bipolar world at the end of the twentieth century, social analysts, economists, and politicians increasingly termed the new age as the age of globalization in which nation states are replaced by transnational global regimes as the most advanced pattern of societal organization. Indeed, the expanding networks of electronic communication, the growth of international trade, the rise of transnational organizations, and the emergence of a global culture seem to document this epochal shift. For most analysts the novelty of globalization is beyond doubt and question.

Many prophets of the new era ignored, however, that there have been global networks also in the premodern world. Even the classical world of the ancient Mediterranean societies was connected by long-distance trade of slaves, luxury goods, and precious metals, caravans traveled between ancient China and the Muslim world, monastic orders established networks that transcended the boundaries of political territories, the early modern colonial empires turned the Atlantic ocean into an interface for the exchange of culture, slaves, and commodities, cultural and intellectual movements like the enlightenment reached out for an international public sphere, etc. Indeed, if we take a historical point of view, "transnational" networks of circulation and communication cannot claim to be a special mark of our time. Warriors and slaves, pilgrims and merchants have always moved across political boundaries. Cultural hybridization and mass

B. Giesen et al., *Human Societies*, Cultural Sociology,
https://doi.org/10.1007/978-3-032-07697-7_9

migration, even foreign investments have been regular phenomena in premodern times. In some respects our world is even less global than previous ones. For example, the amount of international trade in relation to the national one has been larger before World War I than it is today when three major economic regions are drifting apart. So what is new about our contemporary globalization?

The first innovation that makes a difference between today's global circulation of goods, people, and information and the traditional modes of global exchange is the enormous acceleration of circulation and communication. When no roads or ships are available it takes years to walk from one continent to another. Even after the invention of offshore sailing, horse riding, and road construction, long-distance traveling remained a time and energy consuming matter. Traveling the silk route between Beijing and Venice or Istanbul took a year on horseback or camel, today it takes twelve hours by jet plane, sending an administrative report from Lima to Madrid in the Spanish colonial empire required several months even by fast sailing ships, today an email can transfer it within seconds. Moving people, goods, and information in premodern times was so inconvenient and often so dangerous that most people traveled only for sheer necessity and serious reasons—they never did this for pleasure. Furthermore the costs of transporting goods and information was usually so high that only the most precious of goods and the most important pieces of information were worth it. Today transportations increase in speed and its decrease of costs facilitates circulation and communication, and turns traveling into a relatively comfortable activity, even into a pleasure that is accessible for large parts of the population in Western societies. Tourism as a leisure activity was reserved to the aristocracy still in the eighteenth century. It was not before the twentieth century that ordinary people could consider long-distance traveling a regular part of their holidays. Today, millions of tourists migrate periodically to the favorite vacation areas to enjoy safe "adventures," "pristine" nature, and "authentic" culture.

Similar changes occurred on the levels of goods and commodities. Till the last decades of the last century goods for daily consumption—food in particular—were mostly produced in the region or on the national territory where they were consumed. Luxury goods, machinery, and resources like coal or oil were traded on international markets, but there was no German beer, Australian wine, or African fish in American stores. Transportation costs and the risk of perishing during a long journey were too high. Today, however, the world's delicacies are available for our table at modest cost.

The ease of importing exotic food engenders and also impacts our eating habits and styles. Food and diet were, till the last century, in most cases locally produced and ethnically rooted. Immigration and globalization decoupled food and diet from its embeddedness in local and ethnic communities. Pizza and Sushi, Hamburger and Dim Sum are global dishes that can be consumed almost by everyone, everywhere, anytime.

The speed of global circulation and communication also changes our perception of local and cultural differences. Whoever guided a caravan of camels or a mail coach from one travel post to the next to hand it over to the next guide was actually part of a global network but he never moved to a distant territory and he never encountered a different culture. Instead he remained within his familiar localities like the ground staff of airports in distinction to pilots. Even the traveling merchant, the sailor, or the missionary in premodern times did not experience a rapid variation of their environment. The slow speed of transportation allowed only for a gradual change. After months of inconvenient traveling they arrived at their destination and there was usually no return for a long time. Today, by contrast, we leave the plane after an intercontinental flight and we experience the difference in daytime, temperature, and language without this gradual transition. Change and variation occur right in front of us and we can hardly ignore it. But if we get accustomed to change and variation we also are less frightened and intimidated by it. Encountering strangers and coping with cultural differences becomes part of our daily routines or even a pleasant leisure activity.

Tourism is one of the fields in which this pleasant encounter with cultural difference is institutionalized. Tourists enjoy cultural differences as exotic entertainment without understanding the life-world of the local people they visit and take photos of. If, in the process of understanding, they would step into the life-world of the locals, then these local people would get rid of their fascinating exotic aura, the locals would be ordinary and banal, not worth a photo. The lure of the extraordinary that catches the attention of tourists presupposes that we are not familiar with it. But the encounter with the extraordinary has also to be safe and without any risk. Tourists want to hunt warranted trophies without the danger of being attacked by wild animals or savages. Other institutions that allow for the pleasant thrill of a safe encounter with tamed aliens or weird objects are museums or zoos.

The contemporary focus on globalization does not only refer to an increased awareness of global networks of circulation. It is also related to the rise of transnational organizations. Transnational organizations like the UN, the EU, OPEC, or multinational corporations not only evade the range of national politics but they, on their turn, impact strongly on the internal politics of nation states. Here, too, we have to admit that transnational organizations are hardly a historical novelty. From city federations in ancient Greece, via universal churches and monastic orders in the medieval age, to intercontinental colonial empires and city leagues like the Hanseatic League in early modern times, the international Red Cross and the League of Nations after World War I—there have been manifold varieties of transnational organizations before our area. The three or four centuries in which the prince or the nation could claim absolute sovereignty are rather a historical exception than the regular case.

Transnational organizations may not be a new institutional pattern, but the dynamics of many modern supranational or transnational organizations differ from premodern transnationals. Churches and monastic orders, city leagues and colonial empires were driven by the thrust to expand at the cost of competitors or to defend a common cause against an enemy. They could exist and survive even if large parts of the globe were beyond their control. Global organizations like the UN, WHO, IMF, or WTO, by contrast, try to turn all potential members into actual members and their power increases to the degree they succeed in doing so. They are corporations that claim to represent all possible members and this claim is usually unchallenged. Membership in these inclusive corporations is not coercive, but it comes close to it. In this respect they can be compared to states that impose their laws on all persons on their territory irrespective of citizenship. As nobody can refuse to comply with the law even if she is not a citizen of the respective state a nation state can hardly escape from UN intervention because it is not a member of the UN. Imposing a global normative order on deviant states presupposes that a clear majority of possible members is supporting a supranational organization. From the point of view of these inclusive supranational corporations there is no space for international politics any more. Politics consists of intervention and support, of policing and aiding on a worldwide stage. Thus sovereignty seems to be gradually shifted from the nation state to the supranational corporation.

But it is not only on the level of political sovereignty that the new supranational pattern spreads. Other supranational organizations like, for example, the International Sociological Association, the IMF, or the WHO claim, in a similar way, to represent exclusively all possible members with respect to their functional domain—matters of sociology, currency, or health.

Other transnationals like the WWF, Red Cross, Greenpeace, or other nongovernmental organizations are still based on voluntary membership. They cannot impose rules by force but they publicly stigmatize actions that do not conform to assumed rules. They act in a worldwide public sphere and try to mobilize a global outrage. Finally there are, of course, also transnationals that compete with each other in the attempt to control certain fields in a way similar to premodern transnationals. The big multinational corporations are cases in point. They escape the control by national governments because they can withdraw their production sites from one state to another. This volatility of capital and investment is, however, no longer limited to big corporations and international financial markets. Today middle-sized enterprises, too, shift their production from countries where labor cost and taxes are high to countries offering better conditions. Again the decreasing importance of transportation costs facilitates this mobility.

Transnational corporations as well as NGOs act beyond control by nation states but they have to face an emerging global public sphere in which collective identities are staged, represented, and defended. Television pictures that are broadcasted across the boundaries of national or linguistic communities can mobilize compassion and outrage of a global scale. People in the nineteenth century did not yet show this empathy with distant suffering. Solidarity was tied to kinship and locality, class and nationality. Today media reports about catastrophes, civil wars or natural disasters trigger off global movements of solidarity with anonymous victims. This solidarity is certainly rooted in the Western world but it is not confined to it.

But the emergence of a global public sphere does not only facilitate empathy and compassion with anonymous suffering. It also increases sensitivities with respect to offenses that in previous times passed unnoticed because they occurred in a foreign country. Today people can feel offended by a public statement, a joke, a caricature abroad because they know that the message about this statement will be broadcasted internationally. The offense does not primarily result from the fact that somebody dares to say it, but, instead, from the fact that a global public audience can read it, see it, or hear it. What, in one country, was thought to be funny and witty is, in another country, seen as a matter of shame- and honor-provoking counterattacks and acts of revenge and no legitimate authority can settle the issue. Thus the global public sphere fosters not only the collective identity of humankind, but it engenders also new divides and conflicts that are, however, internal affairs of the emerging global society.

The dynamics of culture, too, gets a global turn. Symbols and icons, movies and tunes are increasingly uprooted from their local origins and translated into a global cultural language. The signs of Coca-Cola and Mercedes Benz, the Ode of Joy from Beethoven and the songs of the Beatles, Leonardo's *Mona Lisa* and Michelangelo's *David*, Levi's Jeans and Gucci bags, pizza and beer, chairman Mao, or Stars and Stripes flags are part of this global culture. Almost everyone knows the meaning of these cultural symbols. Sometimes, as in the case of Coca Cola or McDonald's, the sign or name of the particular brand or piece of art is even blended into a generic category.

But cultural globalization does not only amount to an emerging global sphere of symbols that engenders homogeneity and blurs cultural differences and local variation. It also increases variation and allows for new cultural differences. When, for example, Indian immigrants living in Brooklyn can watch the latest movies from Mumbai on TV or Pakistanis living in London can get the news from Al Jazeera every day, the modern media of telecommunication provide a new basis for the conservation of cultural ties between the Diaspora of immigrant community and their culture of origin. It constructs boundaries between the immigrant community and the surrounding majority. Thus global cultural transfer and reproduction results not only in the widespread diffusion of the same cultural pattern, but also in the coexistence and proximity of different cultural patterns in one locality, not only in homogenization, but also in new variation and difference. Different cultural patterns coexisting in a lifeworld can also be blended or merged into a new one. The fusion of food, music, fashion styles, languages, or art from different cultural origins creates new hybrid forms that, after some time, cease to hint at their original components. This "creolization" of culture is as important a process of cultural globalization as the diffusion of global icons.

The process of globalization does not proceed uncontested and unchallenged. There is and, since the nineteenth century, there has always been a strong resistance to the impersonal rule of money and markets. Today's anti-globalization movement is just the latest version of a series of social movements in defense of natural and just conditions of human life against "inhumane," "cold," and "soulless" modernity. Interestingly, the anti-globalization movement has to adapt to the structure of its target—it has to operate transnationally. Participants at the big meetings like Porto Alegre come from all continents to rally against issues all over the globe.

It would be misleading to assume that globalization is an irresistible force that will sweep away all evolutionary earlier institutions and cultural patterns. The nation state is not doomed to fade away and to be replaced by global regimes. It is the nation state that disposes of armies, sets economic and social policies, establishes barriers to foreign imports, mobilizes collective identities for sport and war, decides about the support of cultural activities, and it is the sovereign nation state that movements of independence in Palestine, Kashmir, Kurdistan, Kosovo, Scotland, Sudan etc. thrust for. Far from being replaced by global regimes the nation state is still the most significant arena of democratic representation and the prime warrant for the enforcement of laws. No internationally approved law, contract, or agreement can dispense from being enacted or ratified by national governments. Transnational corporations, nongovernmental organizations, and multinational corporations may narrow the range of political maneuver of national governments but the claim of national sovereignty is hardly less legitimate because of increasing international interdependencies.

Furthermore, the rise of the Internet as a mode of global communication and the growing awareness of a global public sphere have not dissolved the salience of bodily presence in a local site. Not only do seemingly "outdated" modes of oral communication—conferences, lectures, seminars—survive in the age of the Internet and TV media, but locality and bodily presence even get increasing significance and attention. Holiday tourism would collapse if bodily presence in a vacation resort could be substituted by watching a video of it. The experience of the body by physical exercise or by dining, the communication by gestures and bodily presence of conviviality cannot be replaced by a chat in the Internet. People are not content in just watching the big events on TV (e.g., Pope's funeral, rock concerts, sport matches) but, instead, many of them take the inconveniences of traveling to the local site although the TV presentation provides a much closer look at the event. In distinction from the virtual world in which everything is just a convenient mouse click away, being bodily present at the local site of the event gets a new fascination and attraction.

The impact of globalization does not extend to everyone and everywhere. While most people in the Western world, in Japan, and Korea use mobile phones and the Internet, can see Hollywood serials on TV, and know about the effects of currency exchange rates, there are large populations that are decoupled from global networks of communication and circulation. For lack of access to the media or for sheer poverty they are exempted from the beneficial as well as from the detrimental effects of globalization. The life-worlds of homeless people in New York, of peasants in a remote Indian hammock, of a family in central Kongo or of rag pickers in Karachi center much more elementary problems such as the daily search for food or the protection against violence. For them not even the conditions of life in early modern territorial states apply. Seen from the point of view of global regimes they do not fit into the social processes of production, circulation, and communication. In a way they are regarded as superfluous—just human bodies in need of basic protection and food. Invisible for markets and bureaucracies, public discourse and industrial production, they get our attention only in extraordinary moments—when they are hit by catastrophes, when their bare life is at stake, when children die, and when they come too close in an attempt to escape their misery. To keep them at a distance and to get rid of haunting pictures, we establish camps and provide food and medical aid. Enclosures hide those human beings who have just their bare life and for whom we restore just these conditions of bare life. Global regimes, too, have to cope with those who do not fit into its classificatory grid, who cannot be integrated into their modes of production and circulation, and who escape its media of representation.

Bibliography

Alexander, Jeffrey C. (1987): *Twenty Lectures. Sociological Theory Since World War II*. New York: Columbia University Press.

Alexander, C. Jeffrey (2021): "In memoriam Bernhard Giesen (20. Mai 1948 – 26. Dezember 2020)". In: *Soziologie* 50(2).

Alexander, Jeffrey C., Dominik Bartmanski & Bernhard Giesen (2012, ed.): *Iconic Power. Materiality and Meaning in Social Life*. New York; Houndmills: Palgrave.

Alexander, Jeffrey C., Ron Eyerman, Bernhard Giesen, Neil J. Smelser & Piotr Sztompka (2004): *Cultural Trauma and Collective Identity*. Berkeley: University of California Press.

Alexander, Jeffrey C., Bernhard Giesen & Jason L. Mast (2006, ed.): *Social Performance. Symbolic Action, Cultural Pragmatics, and Ritual*. Cambridge: Cambridge University Press.

Alexander, Jeffrey C., Bernhard Giesen, Neil Smelser & Richard Münch (1987, ed.): *The Micro-Macro Link*. Berkeley: University of California Press.

Agamben, Giorgio (1998): *Homo Sacer. Sovereign Power and Bare Life*. Stanford: Stanford University Press.

Aristotle (1967): *Politics*. London: William Heinemann; Cambridge: Harvard University Press.

Bartmanski, Dominik & Ian Woodward (2015): *Vinyl. The Analogue Record in the Digital Age*. London: Bloomsbury.

Baumard, Nicolas, Alexandre Hyafil, Ian Morris & Pascal Boyer (2015): "Increased Affluence Explains the Emergence of Ascetic Wisdoms and Moralizing Religions". In: *Current Biology* 25(1), 10–15.

B. Giesen et al., *Human Societies*, Cultural Sociology,
https://doi.org/10.1007/978-3-032-07697-7

BBC (2016): "Russia's border doesn't end anywhere, Vladimir Putin says". BBC, November 24, 2016; https://www.bbc.com/news/world-europe-38093468

Bellah, Robert N. (2011): *Religion in Human Evolution From the Paleolithic to the Axial Age*. Cambridge; London: Harvard University Press.

Benjamin, Walter (1968): "The Work of Art in the Age of Mechanical Reproduction". In: *Illuminations. Essays and Reflection*. New York: Schocken Books, 217–251.

Bloch, Marc (1973): *The Royal Touch. Sacred Monarchy and Scrofula in England and France*. London: Routledge & Kegan Paul; Montreal: McGill-Queen's University Press.

Boas, Franz (1974): *The Shaping of American Anthropology, 1883–1911: A Franz Boas Reader, edited* by George W. Stocking, Jr. New York: Basic Books.

Burkert, Walter (1998): *Creation of the Sacred. Tracks of Biology in Early Religions*. Cambridge: Harvard University Press.

Castiglione, Baldassare (1900): *The Book of The Courtier. From the Italian of Count Castiglione: Done into English by Sir Thomas Hoby. Anno 1561. With an Introduction by Walter Raleigh*. London: David Nutt in the Strand.

Chagnon, Napoleon A. (1992): *Yanomamö. The Last Days of Eden*. San Diego: Harcourt and Brace.

Childe, Gordon (1951): *Social Evolution*. New York: H. Schuman.

Douglas, Mary (1966): *Purity and Danger. An Analysis of Concepts of Pollution and Taboo*. New York; Washington: Frederick A. Praeger.

Durkheim, Émile (1965): *The Elementary Forms of Religious Life*. New York: Free Press.

Durkheim, Émile (1969): "Individualism and the Intellectuals". In: Lukes, Steven: "Durkheim's 'Individualism and the Intellectuals'". In: *Political Studies. Volume 17(1)*, 19–30.

Eisenstadt, Shmuel N. (1982): "The Axial Age. The Emergence of Transcendental Visions and the Rise of Clerics". In: *European Journal of Sociology / Archives Européennes de Sociologie* 23(2), 294–314.

Eisenstadt, Shmuel N. (1986, ed.): *The Origins and Diversity of Axial Age Civilizations*. Albany: State University of New York Press.

Eisenstadt, Shmuel N. & Bernhard Giesen (1995): "The Construction of Collective Identity". In: *European Journal of Sociology / Archives Européennes de Sociologie / Europäisches Archiv für Soziologie* 36(1), 72–102.

Eliade, Mircea (1959): *The Sacred and the Profane. The Nature of Religion*. New York et al.: Harcourt.

Elias, Norbert (1978): *The Civilizing Process. The Development of Manners*. New York: Urizen Books.

Elias, Norbert (1984): *The Court Society*. New York: Pantheon Books.

Enzensberger, Hans Magnus (1994): *Civil Wars. From L.A. to Bosnia*. New York: The New Press.

Evans-Pritchard, Edward Evan (1976): *Witchcraft, Oracles, and Magic Among the Azande*. Oxford: Oxford University Press.

Fraser, Nancy (2022): *Cannibal Capitalism. How Our System Is Devouring Democracy, Care and the Planet–and What We Can Do about It*. London; New York: Verso.

Gellner, Ernest (1989): *Plough, Sword and Book. The Structure of Human History*. Chicago: University of Chicago Press.

Giesen, Bernhard (1980): *Makrosoziologie. Eine evolutionstheoretische Einführung*. Hamburg: Hoffmann und Campe.

Giesen, Bernhard (1987): "Beyond Reductionism. Four Models of Relating Micro and Macro Levels". In: Alexander, Jeffrey C., Bernhard Giesen, Neil Smelser and Richard Münch (eds.): *The Micro-macro Link*. Berkeley: University of California Press, 337–355.

Giesen, Bernhard (1991a): "Code, Process and Situation in Cultural Selection". In: *Cultural Dynamics* 4(2), 172–185.

Giesen, Bernhard (1991b): *Die Entdinglichung des Sozialen. Eine evolutionstheoretische Perspektive auf die Postmoderne*. Frankfurt/Main: Suhrkamp.

Giesen, Bernhard (1998/1993): *Intellectuals and the German Nation. Collective Identity in an Axial Age*. Cambridge; New York: Cambridge University Press.

Giesen, Bernhard (1999): *Kollektive Identität. Die Intellektuellen und die Nation 2* Frankfurt/Main: Suhrkamp.

Giesen, Bernhard (2004): "Noncontemporaneity, Asynchronicity and Divided Memories". In: *Time Society* 13(1), 27–40.

Giesen, Bernhard (2010): *Zwischenlagen. Das Außerordentliche als Grund der sozialen Wirklichkeit*. Weilerswist: Velbrück.

Giesen, Bernhard (2012): "Inbetweeness and Ambivalence". In: Alexander, Jeffrey C., Ronald N. Jacobs & Philip Smith (eds.): *The Oxford Handbook of Cultural Sociology*. Oxford: Oxford University Press, 788–804.

Girard, René (1979): *Violence and the Sacred*. Baltimore: Johns Hopkins University Press.

Goffman, Erving (1963): *Behavior in Public Places. Notes on the Social Organization of Gatherings*. New York: Free Press of Glencoe.

Goffman, Erving (1971): *Relations in Public. Microstudies of the Public Order*. New York: Basic Books, Inc., Publishers.

Goody, Jack (2001): *The Logic of Writing and the Organization of Society*. Cambridge: Cambridge University Press.

Graeber, David and David Wengrow (2021): *The Dawn of Everything. A New History of Humanity*. London: Penguin.

Habermas, Jürgen (1989): *The Structural Transformation of the Public Sphere. An Inquiry into a Category of Bourgeois Society*. Cambridge, Mass.: The MIT Press.

Harari, Yuval Noah (2014): *Sapiens. A Brief History of Humankind*. London: Random House.

Harner, Michael (1972): *The Jivaro. People of the Sacred Waterfalls*. Berkeley and Los Angeles: University of California Press.

Havel, Václav (2007): *To the Castle and Back. Reflections on My Life as a Fairy-Tale Hero*. New York; Toronto: Alfred A. Knopf.

Jaspers, Karl (1965): *The Origin and Goal of History*. New Haven; London: Yale University Press.

Klein, Ezra (2025): "Trump 2.0 and the Return of 'Court Politics'". *The New York Times*, January 10, 2015; https://www.nytimes.com/2025/01/10/opinion/ezra-klein-podcast-erica-frantz.html

Lee, Richard B. and Irven DeVore (1972, eds.): *Man the Hunter*. Chicago: Aldine Atherton.

Lenski, Gerhard E. and Jean Lenski (1974): *Human Societies. An Introduction to Macrosociology*. New York: McGraw-Hill.

Lévi-Strauss, Claude (1969): *The Elementary Structures of Kinship*. Boston: Beacon Press.

Lévi-Strauss, Claude (1976): *Tristes Tropiques*. Harmondsworth: Penguin Books.

Lovejoy, Arthur Oncken (1966): *The Great Chain of Being. A Study of the History of an Idea*. Cambridge: Harvard University Press.

Malinowski, Bronislaw (1984): *Argonauts of the Western Pacific. An Account of Native Enterprise and Adventure in the Archipelagos of Melanesian New Guinea*. Prospect Heights: Waveland Press.

Malinowski, Bronislaw (1989): "The Problem of Meaning in Primitive Languages". In: Ogden, C. K. and I. A. Richards (eds.): *The Meaning of Meaning. A Study of the Influence of Language upon Thought and of the Science of Symbolism*. San Diego, New York and London: Harcourt Brace Jovanovich Publishers, 296–336.

Mann, Michael (1986): *The Sources of Social Power. Vol. I*. Melbourne: Press Syndicate of the University of Cambridge.

Marx, Karl and Friedrich Engels (2002): *The Communist Manifesto*. London: Penguin Books.

Mauss, Marcel (1990): *The Gift. The Form and Reason for Exchange in Archaic Societies*. New York: W.W. Norton.

Radcliffe-Brown, A. R. (1979): *Structure and Function in Primitive Society. Essays and Addresses*. London and Henley: Routledge & Kegan Paul.

Reckwitz, Andreas & Hartmut Rosa (2023): *Late Modernity in Crisis: Why We Need a Theory of Society*. Cambridge; Medford: Polity.

Rosenberg, Steve (2024): "This morning an electronic billboard on my way to work is displaying this Putin quote: 'Russia's borders do not end anywhere'". Tweet on *X*, January 15, 2024; https://x.com/BBCSteveR/status/1746784252312891463

Sahlins, Marshall D. (1972): *Stone Age Economics*. Chicago: Aldine-Atherton.

Sahlins, Marshall D. & Elman R. Service (1990): *Evolution and Culture*. Ann Arbor: University of Michigan Press.

Schwartz, Barry (1981): *Vertical Classification. A Study in Structuralism and the Sociology of Knowledge*. Chicago: University of Chicago Press.

Service, Elman R. (1975): *Origins of the State and Civilization. The Process of Cultural Evolution*. New York: W.W. Norton & Company.

Service, Elman R. (1979): *The Hunters.* Englewood Cliffs, New Jersey: Prentice-Hall.

Simmel, Georg (1950): "The Stranger". In: Wolff, Kurt H. (ed.): *The Sociology of Georg Simmel.* Glencoe: The Free Press, 402–408.

Simmel, Georg (1990): *The Philosophy of Money.* London and New York: Routledge.

Smith, Adam (1976): *An Inquiry Into the Nature and Causes of The Wealth of Nations.* Chicago: University of Chicago Press.

Speck, Ulrich (2024): "'Russia's borders do not end anywhere': The very definition of empire." Tweet on *X*, January 15, 2024; https://x.com/ulrichspeck/status/1746805116978532659

Troeltsch, Ernst (1931): *The Social Teaching of the Christian Churches. Vol. 1.* London: George Allen & Unwin; New York: The Macmillan Company.

Turner, Victor (1989): *The Ritual Process. Structure and Anti-Structure.* Ithaca: Cornell University Press.

Van Gennep, Arnold (1960): *The Rites of Passage. A Classic Study of Cultural Celebrations.* Chicago: University of Chicago Press.

Veblen, Thorstein (1957): *The Theory of the Leisure Class. An Economic Study of Institutions.* London: George Allen & Unwin.

Weber, Alfred (1997/1935): *Kulturgeschichte als Kultursoziologie.* Marburg: Metropolis.

Weber, Alfred (2012/1920): "Fundamentals of Cultural Sociology: Social Process, Civilizational Process, and Cultural Movement (1920)". In: Loader, Colin (ed.), *Alfred Weber and the Crisis of Culture, 1890–1933.* New York: Palgrave Macmillan, 165–204.

Weber, Alfred (2015/1913): "The Sociological Concept of Culture". In: *Cultural Sociology* 9(2), 258–270.

Weber, Max (1964): *The Theory of Social and Economic Organization.* New York: The Free Press.

Weber, Max (1967a): "Science as a Vocation". In: Gerth, Hans H. and C. Wright Mills (eds.): *From Max Weber: Essays in Sociology.* New York: Oxford University Press, 129–156.

Weber, Max (1967b): "Religious Rejections of the World and Their Directions". In: Gerth, Hans H. and C. Wright Mills (eds.): *From Max Weber: Essays in Sociology.* New York: Oxford University Press, 323–359.

Weber, Max (1978): *Economy and Society. An Outline of Interpretive Sociology.* Berkeley: University of California Press.

Whitehead, Alfred North (1967): *Science and the Modern World.* New York: The Free Press.

Name Index[1]

[1] Note: Page numbers followed by 'n' refer to notes.

B. Giesen et al., *Human Societies*, Cultural Sociology,
https://doi.org/10.1007/978-3-032-07697-7

Subject Index[1]

[1]Note: Page numbers followed by 'n' refer to notes.

B. Giesen et al., *Human Societies*, Cultural Sociology,
https://doi.org/10.1007/978-3-032-07697-7

S

W

9783032076960